AF741299

Biological Research in Aquatic Science

Edited by Yusuf Bozkurt

Published in London, United Kingdom

IntechOpen

Supporting open minds since 2005

Biological Research in Aquatic Science
http://dx.doi.org/10.5772/intechopen.77742
Edited by Yusuf Bozkurt

Contributors
Hamdy Abo-Taleb, Arne Linløkken, Jacky Cosson, Ross F. Tallman, Kevin Hedges, Kimberly L. Howland, Zoya A. Martin, Les N. Harris, Marianne Marcoux, Steven H. Ferguson, Nigel E. Hussey, Jean-Sebastien Moore, Cory J.D. Matthews, Yusuf Bozkurt

© The Editor(s) and the Author(s) 2019
The rights of the editor(s) and the author(s) have been asserted in accordance with the Copyright, Designs and Patents Act 1988. All rights to the book as a whole are reserved by INTECHOPEN LIMITED. The book as a whole (compilation) cannot be reproduced, distributed or used for commercial or non-commercial purposes without INTECHOPEN LIMITED's written permission. Enquiries concerning the use of the book should be directed to INTECHOPEN LIMITED rights and permissions department (permissions@intechopen.com).
Violations are liable to prosecution under the governing Copyright Law.

Individual chapters of this publication are distributed under the terms of the Creative Commons Attribution 3.0 Unported License which permits commercial use, distribution and reproduction of the individual chapters, provided the original author(s) and source publication are appropriately acknowledged. If so indicated, certain images may not be included under the Creative Commons license. In such cases users will need to obtain permission from the license holder to reproduce the material. More details and guidelines concerning content reuse and adaptation can be found at http://www.intechopen.com/copyright-policy.html.

Notice
Statements and opinions expressed in the chapters are these of the individual contributors and not necessarily those of the editors or publisher. No responsibility is accepted for the accuracy of information contained in the published chapters. The publisher assumes no responsibility for any damage or injury to persons or property arising out of the use of any materials, instructions, methods or ideas contained in the book.

First published in London, United Kingdom, 2019 by IntechOpen
IntechOpen is the global imprint of INTECHOPEN LIMITED, registered in England and Wales, registration number: 11086078, The Shard, 25th floor, 32 London Bridge Street
London, SE19SG - United Kingdom
Printed in Croatia

British Library Cataloguing-in-Publication Data
A catalogue record for this book is available from the British Library

Additional hard and PDF copies can be obtained from orders@intechopen.com

Biological Research in Aquatic Science
Edited by Yusuf Bozkurt
p. cm.
Print ISBN 978-1-78923-813-6
Online ISBN 978-1-78923-814-3
eBook (PDF) ISBN 978-1-78923-959-1

For EU product safety concerns:
IN TECH d.o.o., Prolaz Marije Krucifikse Kozulić 3, 51000 Rijeka, Croatia, info@intechopen.com or visit our website at intechopen.com.

We are IntechOpen, the world's leading publisher of Open Access books

Built by scientists, for scientists

4,200+
Open access books available

116,000+
International authors and editors

125M+
Downloads

151
Countries delivered to

Our authors are among the
Top 1%
most cited scientists

12.2%
Contributors from top 500 universities

CLARIVATE ANALYTICS
BOOK CITATION INDEX
INDEXED

WEB OF SCIENCE™

Selection of our books indexed in the Book Citation Index
in Web of Science™ Core Collection (BKCI)

Interested in publishing with us?
Contact book.department@intechopen.com

Numbers displayed above are based on latest data collected.
For more information visit www.intechopen.com

Meet the editor

Yusuf Bozkurt holds BSc, MSc, and PhD degrees from Ankara University (Turkey). He is currently Professor of Biotechnology of Reproduction in the field of Aquaculture at the İskenderun Technical University (Turkey). His research interests include gamete biology and reproductive biotechnology in aquatic species with an emphasis on cryo-conservation.

Contents

Preface

Biological research involves multidisciplinary studies, including field and experimental studies on biological characteristics at the individual level as well as mechanisms influencing organisms' response to internal and external factors. From this point of view, biological research focuses on morphological and physiological characteristics, biotechnology, and molecular mechanisms as well as behavior. Additionally, biological research focuses on interactions among the organisms and their surrounding environment.

Biological Research in Aquatic Science describes biological studies and applications in different research areas. The book is divided into five sections, each containing one chapter written by experts in their respective fields.

Section 1, "Overview on Cryobiology in Aquatic Science," is an introductory section that explains the mechanisms, strategies, and application fields of cryopreservation biotechnology and cryobanking in aquatic science.

Section 2, "Spermatology in Aquatic Species," presents valuable information regarding biological and physiological features of fish sperm as well as modern technologies for the evaluation of its quality.

Section 3, "Climate Change and Aquatic Environment," describes climate change and its effect on fish community, lake ecology, primary production, eutrophication, and water quality in terms of temperate lakes.

Section 4, "Migration in Aquatic Species," covers the migration reasons of aquatic species as well as properties of the Arctic with an emphasis on the Canadian region.

Section 5, "Plankton in Aquatic Ecosystems," describes fish biology in terms of food chain relations and the relationships between fish larva and zooplankton.

This book addresses various topics regarding biology in aquatic science and will be helpful for researchers studying in the field. I would like to thank all the authors for their distinguished contributions, and IntechOpen and its Author Service Manager Ms. Sandra Maljavac for her help in publishing this book.

Yusuf Bozkurt
Professor,
İskenderun Technical University,
İskenderun, Hatay, Turkey

Section 1

Overview on Cryobiology in Aquatic Science

Chapter 1

Introductory Chapter: Cryopreservation Biotechnology in Aquatic Science

Yusuf Bozkurt

1. Introduction

Cryopreservation is the process of freezing the biological materials at temperature of liquid nitrogen (LN_2) (–196°C). This means it is possible storing of the biological materials as unchanged for centuries with the capability of recovering the cell functionality following the thawing process [1].

This situation has been increasing the importance of cryobiology as a science that examines the effect of ultra-low temperatures on cell, tissue, organ, and organisms and also freezability of these structures maintaining their viability. In addition, a better understanding of functional properties of thawed cells following the freezing process has been accelerating the development of cryobiology [2].

The cryopreservation method basically includes temperature reduction, cellular dehydration, freezing, and thawing. The lowering of normal temperature to 4°C reduces the cellular metabolic activity and increases the life span of sperm cells. Following thawing process, normal functions of the cells restart [3]. These events at ultra-low temperatures provide basic mechanisms for long-term preservation of biological material in genetically stable form. In practice, no significant change of biological importance occurs below −150°C, and therefore, the material can be conveniently stored in liquid nitrogen at −196°C [2].

Gamete, embryo, and embryonic cell cryopreservation have become of tremendous value in aquatic biotechnologies, which provide an important tool for the propagation of economically important species, and also in the protection of endangered species and genetic diversity in aquatic species [1].

Following successful cryopreservation of avian spermatozoa using glycerol as cryoprotectant by Polge et al. [4], cryopreservation of male gametes became possible in this research area. For the first time, Blaxter [5] applied a similar approach for fish gametes and reported achieving approximately 80% cellular motility following thawing of Atlantic herring spermatozoa in the field of aquaculture. Since then, cryopreservation of fish sperm has been studied and succeeded in more than 200 species [6, 7]. Today, sperm management techniques have been established for freshwater and marine fish species [8–12].

Cold or frozen state preservation of gametes is an important biotechnological tool for aquatic species conservation and has a great concern for aquaculture. Growing in concern to this biotechnology has led to an increase in the number of studies in this research area [13]. Nowadays, it is possible to use preserved semen in routine reproduction applications in aquaculture practices [14].

IntechOpen

2. Cryopreservation of sperm cells in aquatic species

Sperm cryopreservation is a very valuable tool for the conservation of aquatic species [15]. It is considered as a reliable method for the *ex situ* preservation of biodiversity since it provides opportunity to preserve desired cell samples. In addition, it is possible to reconstruct the original strain, population, or variety following required environmental restoration via this biotechnology [2].

The progressive interest in the application of cryopreservation to aquaculture has revealed how useful this method could be in the management of fish reproduction especially when it is combined with other reproductive technologies such as androgenesis or sex reversal [2].

Two methods can be used for gamete cryopreservation: slow freezing and vitrification. Slow freezing uses low concentrations of cryoprotectants, which are associated with chemical toxicity and osmotic shock. Semen refrigeration with slow cooling rates (0.5–1°C/min) and temperature reduction induces stress on cell membranes and causes modifications in the functional state of membranes [16]. The stress caused by ice crystal formation is associated with the osmotic pressure changes in the unfrozen solution [17].

Cold shock reduces membrane permeability to water and solutes resulting in membrane injury. The main changes occurring during the freezing process are ultrastructural, biochemical, and functional. These changes reduce fertilization of eggs. In frozen/thawed semen, motility of sperm cells is better protected than its morphological integrity. Membrane permeability is increased following cooling process, and this may be a result of increased membrane leakiness and specific protein channels [18].

3. Cryopreservation of eggs and embryos in aquatic species

Cryopreservation of egg is more complicated than that of sperm. It is possible to indicate that the large size and the presence of three different membrane layers with different water permeabilities are the main obstacles related to the removal of intracellular water from fish eggs [2].

When the eggs are stripped, the eggs are permeable to ice-reducing cryoprotectants due to opening of channel in the shell-like chorion. According to Harvey and Ashwood-Smith [19], penetration of cryoprotectants such as glycerol, DMSO, and methanol is rather slow in unactivated ova. However, once fertilization or activation occurs, the channel closes and the chorion hardens [20]. At this point, the eggs are impermeable to cryoprotectants and will significantly increase in size due to a brief inflow of the water [21].

Cryopreservation of embryos has become an integral part of assisted reproduction. Successful cryopreservation of embryos is important because the biodiversity of both paternal and maternal genomes will be preserved. Fish embryos are better candidates than eggs for the cryopreservation process due to their higher membrane permeability, less chilling sensitivity, and less complex membrane system.

Studies carried out so far associated to fish egg cryopreservation have been mainly focused on model species such as zebrafish (*Danio rerio*) [22], although other marine and freshwater species have also been studied, for example, gilthead seabream (*Sparus aurata*) [20] and some South American freshwater species [23]. On the other hand, there are factors limiting fish egg cryopreservation including their multicompartmental biological systems, high chilling sensitivity, low membrane permeability, and larger size [20].

4. Cryobanking in aquatic science

One of the important fields using cryopreservation technology is the cryobanks or sperm banks. Cryobanks are currently more developed for rare domestic animals such as cattle, sheep, and goats than for non-domestic animals. In addition, use of cryobanking to facilitate the management and conservation of endangered species is becoming widespread [24].

The creation of cryobanks for the selected stock to prevent outbreaks or genetic drift is also essential to develop genetic selection programs in commercial aquaculture. On the other hand, conservation of aquatic species which in danger of extinction, is also necessary until the environmental conditions recovered.

In aquatic science, cryobanking has considerable advantages on cultured aquatic species in captivity, in terms of cost, labor, and security since thousands of samples from different generations can be maintained in a minimum space without the risk of loss caused by disease [25]. Moreover, transportation and management of frozen samples are relatively simple, allowing greater flexibility for designing recovery programs. In addition, development of reproductive technologies in aquatic species allows recovery of population from semen samples through cross-breeding programs or application of androgenesis procedures [26].

Research on fish germplasm cryobanking has been carried out on different cell types such as sperm cells, somatic cells, spermatogonia and primordial germ cells, as well as oocytes and embryos. On the other hand, it is well known that sperm cells present advantages compared to other cell types because of their small size and high resistance to chilling [27].

5. Opportunities and new strategies

The cryopreservation process provides many benefits. Some of the cryobiological applications in the field of aquaculture have been summarized as follows ([6, 8, 28]:

- Storing the sperm for routine fertilization process
- Cross-breeding programming independently the maturation period or availability of breeders
- Utilizing all the sperms from species with a large production
- Increasing the fertile life of the individuals
- Transporting the gametes or embryos between farms instead of breeders
- Marketing the well-characterized and standard quality sperm
- Hybridizing between species with different maturation periods
- Reducing the synchronization treatments
- Year-round supplying the broodstock gametes

Recently, new technologies have been developed to conserve paternal and maternal genetic information. From this point of view, latest studies have focused on cryopreservation of primordial germ cells as an alternative for the cryopreservation of both paternal and maternal genomes. On the other hand, there is little data about reprogramming of somatic cells into primordial germ cells in fish. In addition, cryopreservation of fish tissues can be considered for cryobanking. However, the regeneration methods should be well studied and established in aquatic species [2].

6. Conclusion

Gamete, embryo, and embryonic cell cryopreservation have become of tremendous value in aquatic biotechnologies, which provide an important tool for the propagation of economically important species, and also in the protection of the endangered species and genetic diversity in aquatic species [1]. This situation has been increasing the importance of cryobiology as a science, examining the effect of ultra-low temperatures on cells, tissues, organs, and organisms and also the freezability of these structures maintaining their viability [2].

Author details

Yusuf Bozkurt
Faculty of Marine Sciences and Technology, İskenderun Technical University, İskenderun, Hatay, Turkey

*Address all correspondence to: yfbozkurt@hotmail.com

IntechOpen

© 2019 The Author(s). Licensee IntechOpen. This chapter is distributed under the terms of the Creative Commons Attribution License (http://creativecommons.org/licenses/by/3.0), which permits unrestricted use, distribution, and reproduction in any medium, provided the original work is properly cited.

References

[1] Bozkurt Y. Intyroductory chapter: Application fields of cryopreservation biotechnology. In: Bozkurt Y, editor. Cryopreservation Biotechnology in Biomedical and Biological Sciences. London, UK: IntechOpen Publishing; 2018. pp. 1-4. ISBN: 978-1-78984-680-5

[2] Bozkurt Y. The Role of Cryobiology in Conservation of Aquatic Genetic Resources. Saarbrücken, Germany: Lambert Academic Publishing; 2017. p. 94. ISBN: 978-3-330-05346-5

[3] Mazur P. Freezing of living cells: Mechanisms and implications. The American Journal of Physiology. 1984;**247**:125-142

[4] Polge C, Smith AU, Parkes AS. Revival of spermatozoa after vitrification and dehydration at low temperatures. Nature. 1949;**164**(4172):666

[5] Blaxter JHS. Sperm storage and cross-fertilization of spring and autumn spawning herring. Nature. 1953;**172**:1189-1190

[6] Tiersch TR, Yang H, Jenkins JA, Dong Q. Sperm cryopreservation in fish and shellfish. Society of Reproduction and Fertility Supplement. 2007;**65**:493-508

[7] Tsai S, Spikings E, Lin C. Effects of the controlled slow cooling procedure on freezing parameters and ultrastructural morphology of Taiwan shoveljaw carp (*Varicorhinus barbatulus*) sperm. Aquatic Living Resources. 2010;**23**:119-124

[8] Suquet M, Dreanno C, Fauvel C, Cosson J, Billard R. Cryopreservation of sperm in marine fish. Aquaculture Research. 2000;**31**(3):231-243

[9] Van der Straten KM, Leung LK, Rossini R, Johnston SD. Cryopreservation of spermatozoa of black marlin, *Makaira indica* (*Teleostei: Istiophoridae*). CryoLetters. 2006;**27**(4):203-209

[10] Yavas I, Bozkurt Y. Effect of different thawing rates on motility and fertilizing capacity of cryopreserved grass carp (*Ctenopharyngodon idella*) sperm. Biotechnology and Biotechnological Equipment. 2011;**25**(1):2254-2257

[11] Bozkurt Y, Yavas I, Yıldız C. Effect of different avian egg yolk types on fertilization ability of cryopreserved common carp (*Cyprinus carpio*) spermatozoa. Aquaculture International. 2014;**22**:131-139

[12] Bozkurt Y, Yavas I, Bucak MN, Yeni D. Effect of different cryoprotectants (Gly, Methanol and Dimethylsulfoxide) on post-thaw quality, viability, fertilization ability and DNA damage of cryopreserved Nile tilapia (*Oreochromis niloticus*) spermatozoa. CryoLetters. 2019;**40**(1):11-17

[13] Carolsfeld J, Harvey B, Godinho HP, Zaniboni-Filho E. Cryopreservation of sperm in Brazilian migratory fish conservation. Journal of Fish Biology. 2003;**63**:472-481

[14] Maria A, Carneiro P. Fish semen cryopreservation in Brazil: State of the art and future perspectives. Ciencia Animal. 2012;**22**(1):124-131

[15] Tsai S, Lin C. Advantages and applications of cryopreservation in fisheries science. Brazilian Archives of Biology and Technology. 2012;**55**(3):425-433

[16] Watson PF. The causes of reduced fertility with cryopreserved semen. Animal Reproduction Science. 2000;**60-61**:481-492

[17] Medeiros CMO, Forell F, Oliveira ATD, Rodrigues JL. Current Status of Sperm Cryopreservation: Why Isn't It Better? Theriogenology. 2002;**57**:327-344

[18] Purdy PH. A review on goat sperm cryopreservation. Small Ruminant Research. 2006;**6**:215-225

[19] Harvey B, Ashwood-Smith MJ. Cryopretectant penetration and supercooling in the eggs of samonid fishes. Cryobiology. 1982;**19**:29-40

[20] Zhang T, Rawson DM. Cryo-conservation in fish: The potentials and challenges. Cryobiology. 2007;**55**(3):354

[21] Wallace RA, Selman K. Cellular and dynamic aspects of oocyte growth in Teleosts. American Zoologist. 1981;**21**:325-343

[22] Guan M, Rawson DM, Zhang T. Cryopreservation of zebrafish (*Danio rerio*) oocytes using improved controlled slow cooling protocols. Cryobiology. 2008;**56**:204-208

[23] Streit DP Jr, Godoy LC, Ribeiro RP, Fornari DC, Digmayer M, Zhang T. Cryopreservation of embryos and oocytes of South American fish species. In: Yamashiro H, editor. Recent Advances in Cryopreservation. London, UK: IntechOpen Publishing; 2014. pp. 45-58. ISBN: 978-953-51-1644-8

[24] Wildt DE, Rall WF, Crister JK, Monfort SL, Seal US. Genome resource banks: 'living collections' for biodiversity conservation. Bioscience. 1997;**47**:689-698

[25] Martinez-Paramo S, Perez-Cerezales S, Gomez-Romano F, Blanco G, Sanchez JA, Herraez MP. Cryobanking as tool for conservation of biodiversity: Effect of brown trout sperm cryopreservation on the male genetic potential. Theriogenology. 2009;**71**:594-604

[26] Babiak I, Dobosz S, Goryczko K, Kuzminski H, Brzuzan P, Ciesielski S. Androgenesis in rainbow trout using cryopreserved spermatozoa: The effect of processing and biological factors. Theriogenology. 2002;**57**:1229-1249

[27] Martínez-Páramo S, Horváth A, Labbe C, Zhang T, Robles V, Herráez P, et al. Cryobanking of aquatic species. Aquaculture. 2017;**472**:156-177

[28] Gwo JC. Cryopreservation of sperm of some marine fishes. In: Cryopreservation in Aquatic Species. Edited by Tiersch TR, and Green CC. 2nd edn. 2011, 459-481. World Aquaculture Society, Baton Rouge, Lousiana

Section 2

Spermatology in Aquatic Species

Chapter 2

Fish Sperm Physiology: Structure, Factors Regulating Motility, and Motility Evaluation

Jacky Cosson

Abstract

For reproduction, most fish species adopt external fertilization: their spermatozoa are delivered in the external milieu (marine- or freshwater) that represents both a drastic environment and a source of signals that control the motility function. This chapter is an updated overview of the signaling pathways going from external signals such as osmolarity and ionic concentration and their membrane reception to their transduction through the membrane and their final reception at the flagellar axoneme level. Additional factors such as energy management will be addressed as they constitute a limiting factor of the motility period of fish spermatozoa. Modern technologies used nowadays for quantitative description of fish sperm flagella in movement will be briefly described as they are more and more needed for prediction of the quality of sperm used for artificial propagation of many fish species used in aquaculture. The chapter will present some applications of these technologies and the information to which they allow access in some aquaculture species.

Keywords: flagellum, sperm energetics, sperm signaling, sperm motility, osmolarity

1. Introduction

The main function of a spermatozoon is to convey the male genome remotely to the female one, which occurs in case of fish by swimming in the external milieu, marine or freshwater. Spermatozoa must access, bind, and penetrate an egg, for successful fertilization. Therefore, most of the physiological activity of fish spermatozoa is motility oriented. These processes include as a prerequisite the activation of spermatozoon motility. In case of fish species, spermatozoa stored in the seminal plasma are immotile during transit through the genital tract of most externally fertilizing teleost and chondrostean. Motility is induced immediately following the release of spermatozoa from the male genital tract into the aqueous environment. External trigger agents for the initiation of motility depend on the species' reproductive behavior that is mostly controlled by the aquatic environment (fresh or salt water). Triggering signals include osmotic pressure, ionic and gaseous components of the external media, and, in some cases, egg-derived substances used for sperm guidance. Environmental factors influencing fish spermatozoon motility have received a large attention: these extensive studies led to several mechanisms of activation for freshwater and marine fish spermatozoa. However, after reception of the signal, a transduction pathway initiated by these mechanisms must lead the information to the flagellar motility apparatus (axoneme). This review presents

IntechOpen

the current knowledge with respect to (1) membrane reception of the activation signal and its transduction through the spermatozoon plasma membrane via the external membrane components such as ion channels or aquaporins; (2) cytoplasmic trafficking of the activation signal; (3) final steps of the signaling, including signal transduction to the axonemal machinery, and activation of axonemal dynein motors and regulation of their activity; (4) signaling involved in guidance processes that control the sperm/egg approach and meeting; and (5) pathways supplying energy for the short flagellar motility period of fish spermatozoa. For each step in this signaling process, quantitative methods were developed to evaluate the quality of the sperm samples that are used for aquaculture propagation of many fish species. These methods as well as examples of their usefulness for application to fish artificial reproduction are presented in the last part of this chapter.

2. Fish sperm structure and spermatogenesis

The main traits of the flagellar mechanics of spermatozoa and the main factors that regulate their motility have been described in details in a recent review chapter by Cosson et al. [1].

2.1 Structure of fish spermatozoa: a brief presentation

Compared to mammalian spermatozoa, the structure of a fish spermatozoon is qualified as "simple sperm", mostly because the flagellum structure does not include additional columns flanking the motor part (axoneme) that are present in mammal sperm. In teleost fish, spermatozoa generally have no acrosome (in contrast to chondrostean such as sturgeon), and the impenetrable chorion presents a micropyle that gives access to the membrane of the oocyte. Spermatozoa often show a spherical nucleus with homogenous, highly condensed chromatin, a nuclear fossa, a midpiece of variable size with or without a cytoplasmic channel, and one or two long flagella [2]. Moreover, fish spermatozoa can be classified into two forms, aqua sperm and intro-sperm, according the external or internal mode of fertilization, respectively [2].

The main components present in a fish spermatozoon [3, 4] are: the head is occupied mostly by the nucleus with paternal DNA material. In most fish species, the head has an almost spherical shape (diameter of 2–4 μm). In some case such as sturgeon, paddlefish, and eel spermatozoa, the shape of head is elongated (up to 9 μm long and 2 μm wide) [5–7], the midpiece is mostly composed of the centrioles, and the mitochondria (usually from 2 to 9 per each spermatozoon) is generating energy (ATP) for motility [8]. In a mature spermatozoon, because the protein synthesis machinery is absent, no gene expression occurs. The centriolar complex of midpiece consists of the proximal and the distal centrioles, which forms the basal body of the flagellum, used for anchoring the flagellum to the head of the sperm cell.

The flagellum is a highly conserved organelle during evolution, and there are very few differences between molecular composition of sperm flagella relatively to that in protists [1, 10]. Fish sperm flagellar length varies from 20 to 100 μm, depending on species. Flagellar bending is generated by a highly organized cylindrical system of microtubules, called the axoneme, emanating from the basal body [11]. In turn, the canonical "9 + 2" axoneme consists of nine pairs of peripheral microtubular doublets and one central pair of singlet microtubules. This structural arrangement [12] is illustrated in **Figure 1**. Such axonemal pattern is highly conserved and almost identical among eukaryotic cilia and flagella from protozoans to human. Nevertheless, in Anguilliformes and Elopiformes sperm flagella present a "9 + 0" pattern lacking central microtubules [6, 7, 13].

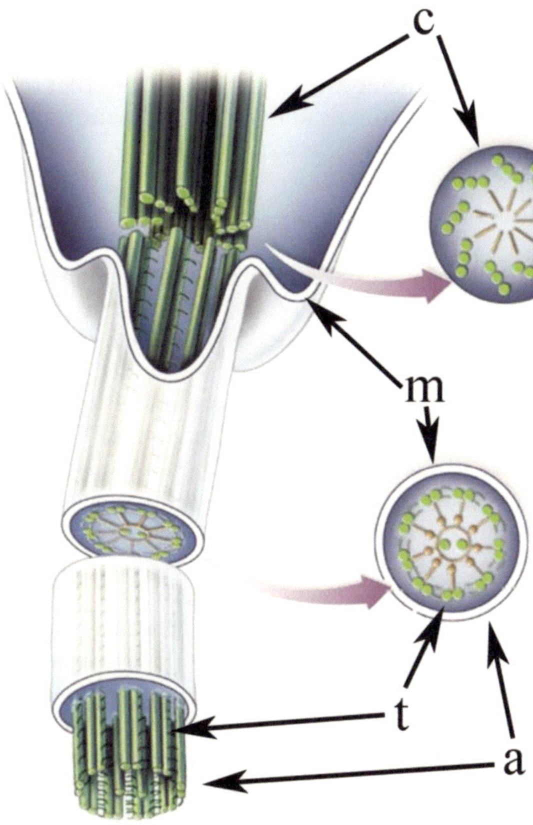

Figure 1.
Schematic structure of the head-tail junction of a model spermatozoon: the artistic view applies to a teleost such as trout [9] as an example of sperm cell. m, membrane surrounding both the head (top) and the tail; a, axoneme, the mechanical part actuating the flagellum (bottom); t, microtubule doublets, the major scaffold of the 9 + 2 axoneme; c, centriolar complex, at the basal part of the axoneme, made of microtubule triplets. In many fish species, the mitochondria (not shown for simplification) are localized in this head-tail junction zone, and two centrioles are orthogonally assembled to form the centriolar complex.

The structural connections between the nine peripheral outer doublets and the sheath surrounding the central pair occur through the radial spokes. The central pair of singlets is enclosed in this sheath of proteins forming a series of projections that are well positioned to interact with each of the spoke heads and regulate the wave propagation [14]. Each of the outer doublets is connected to adjacent pairs of doublets by nexin links, presenting elastic properties allowing to resist the free sliding of the microtubules; nexin is a dynein regulatory protein [15]. The peripheral doublets are strung with two rows of dynein arms along the entire length of microtubules. These dynein arms consist of macromolecular ATPase complex [16, 17] and represent the basic motor actuating the whole axoneme; they extend from an outer doublet toward an adjacent doublet [18]. Both the spokes and the dynein complex contain different calcium-binding proteins so as for flagella to be able to respond to regulation by free calcium concentration through altering their beating pattern [19, 20]. As briefly described above, axonemes are complex structures composed of at least 500 different protein components [21].

The bending process in an axoneme is caused by sliding between two adjacent doublets of outer microtubules that slide relatively to each other due to the motive force, generated by molecular dynein motor activity [16]. Due to enzymatic hydrolysis of ATP by the latter, which induces force generation of the power stroke of individual dynein molecules, the dynein arms interact with tubulin of the B tubule from the adjacent doublet, causing a process of active sliding in a cooperative way [22]. Several local bending processes occur because this sliding activity is present in only some segments of the axoneme at a given time, while other segments remain inactive [4]. Wave propagation from head to tip provokes the translation of the whole spermatozoon in the opposite (forward) direction.

2.2 Spermatogenesis in fish

For obtainment of full efficiency of motility, all the elements of the flagellum must have been, during spermatogenesis, correctly assembled mostly as an elongated structure called axoneme playing the role of propelling engine, surrounded by the plasma membrane, and this device must be provided in energy in terms of ATP [8], the fuel common to many cell types that is mostly generated by mitochondrial respiration in case of fish sperm as detailed in paragraph 4.

Spermatogenesis is an important phase in case of fish spermatozoa because it is the ATP store that constitutes the main source of energy that will sustain the short but highly energy-demanding motility period [8].

A detailed description of fish spermatogenesis is well documented in many species. Briefly, spermatogenesis is a developmental process during which a small number of diploid stem cells (spermatogonia) produce a large number of highly differentiated spermatozoa carrying a haploid, recombined genome, and a structurally complete flagellum. Survival and development of those germ cells depend on their close contact with specific cells called Sertoli cells. Apart from their phagocytic role, the Sertoli cells change the growth factor expression, and subsequently, modulate germ cell proliferation/differentiation via complex mechanisms involving, in fish, both pituitary gonadotropins LH and FSH that stimulate gonadal sex steroid hormone production directly by the activation of another cell type called Leydig cells.

Fishes represent the largest and most diverse group of vertebrates. However, our knowledge on spermatogenesis in this group is limited to a few species used in basic research and/or in aquaculture biotechnologies such as guppy, catfish, cod, eel, medaka, salmon, tilapia, trout, and zebrafish. In an amniote vertebrates (fishes and amphibians), one observes a cystic type of spermatogenesis, which presents two main differences compared to higher vertebrates [23]. First, within the spermatogenic tubules, cytoplasmic extensions of Sertoli cells form cysts that envelope a single, clonally and hence synchronously developing group of germ cells deriving from a single spermatogonium. Second, the cyst-forming Sertoli cells retain their capacity to proliferate also in the adult fish. Sertoli cells are surrounding and nursing one synchronously developing germ cell clone. Different clones being in different stages of development generate a tubular compartment containing differently sized groups of germ cells in different stages of spermatogenesis.

So as for the distribution of spermatogonia in the germinal compartment, one can observe either a first type of restricted spermatogonial distribution, which is found in the higher teleost groups, such as in the order Atheriniformes, Cyprinodontiformes, and Beloniformes, where the distal regions of the germinal compartment are occupied by Sertoli cells surrounding early, undifferentiated spermatogonia. While the cells divide and enter in meiosis and the cysts migrate toward the region of the spermatic ducts located centrally in the testis, this is where spermiation occurs, i.e. the cysts open to release spermatozoa.

In a second type, where an unrestricted spermatogonial distribution that is considered a more primitive pattern found in less evolved taxonomic groups, such as in the order Cypriniformes, Characiformes, and Salmoniformes, occurs, spermatogonia are spread along the germinal compartment throughout the testis. The cysts do not migrate during their development. In addition, intermediate forms also exist between restricted and unrestricted spermatogonial distribution, such as in Perciformes, tilapia, Pleuronectiformes, or Gadiformes.

Therefore, the development of spermatogenic cells strictly depends on their interaction with the somatic elements of the testis, among which Sertoli cells play a crucial role. During fish spermatogenesis, Sertoli cells are formed by mitosis just in time and in exactly the number required. This tailored Sertoli cell proliferation was first described in the guppy [24]. So far, we observe that spermatogenesis is a highly organized and coordinated process, in which diploid spermatogonia proliferate and differentiate to form spermatozoa in their final morphology. The duration of this process is usually shorter in fish than in mammals. In principle, this process can be divided, from a morpho-functional point of view, in three different phases: (i) the mitotic or spermatogonial phase with the different generations of spermatogonia, (ii) the meiotic phase with the primary and secondary spermatocytes, and (iii) the spermiogenic phase with the haploid spermatids emerging from meiosis and differentiating, without further proliferation, into flagellated spermatozoa.

Analysis of the role of hormones reveals a complex process. Three steps at which reproductive hormones play a critical regulatory role are (i) the balance between self-renewal and differentiation of spermatogonial stem cells, (ii) the transition from type A spermatogonia to rapidly proliferating type B spermatogonia, and (iii) the entry into meiosis. During later developmental stages, on the other hand, the endocrine system seems to ensure a permissive rather than stimulatory role, enabling Sertoli cells and possibly other somatic cells to generate a microenvironment that germ cells require to proceed through meiosis and spermiogenesis [25]. In case of fish, three types of spermiogenesis have been described, based on the orientation of the flagellum relatively to the nucleus and on whether or not a nuclear rotation occurs.

2.3 Flagellum genesis

The main organelle in the flagellum, the axoneme, is resulting from the progressive assembly of groups of elements synthesized in the cell body and then transported to the flagellar compartment and delivered to the flagellar tip to elongate it thanks to the motility of specific transporters called intra-flagellar transport (IFT) along the internal side of the flagellar membrane [21]. The trafficking is ensured by two important molecular motors, belonging to the dynein group (retrograde, meaning from tip to base of the flagellum) and to the kinesin group (anterograde, meaning from the cell body to the flagellum tip) [26].

3. Successive steps leading to fish sperm motility

3.1 The maturation step

Maturation is the step following the end of spermatogenesis and that provides to the spermatozoon its ability to respond to motility-activating factors [27]. Signals for maturation are quite various among fish species. Sperm maturation is also regulated by the endocrine system.

Examples of sperm maturation have been studied in details in different fish species: salmonids, cyprinids, and sturgeons. In salmonids, the group of Morisawa, in Japan, demonstrated, in particular in case of salmon, that maturation is mainly under control of cAMP and pH of the water where fish are transiting during migration [28]. In carp, results from Redondo et al. [29] indicate that an ionic equilibration across the sperm membrane is the main factor responsible for maturation. In case of sturgeon species, it was shown that spermatozoa are not able to become activated at simple contact with external water [30]; sperm cells need a transient contact with urine, the latter getting mixed with milt prior to ejaculation [31] which is enough to render spermatozoa fully motile.

3.2 The activation step per se

This is a very brief event lasting a fraction of a second [32, 33]. Among several kinds of activating signals, one can mention (1) osmolarity, the most common, (2) specific ions such as K^+ in a few species (salmonids and sturgeons as examples), and (3) other signaling molecules such as CO_2 in few cases [34]. The osmolarity signal depends on the external medium where fish delivers its sperm: marine fish spermatozoa activate when coming in contact with sea water, a salty solution of high osmolarity, while freshwater fish species shed their sperm in a very low-osmolarity medium. In both cases, the spermatozoa sustain a large stress that consists in a jump from seminal fluid ranging an osmolarity 300 mOsmol/kg to either sea water (around 1100 mOsmol/kg) or freshwater (maximum 50 mOsmol/kg).The environmental osmotic pressure appears consistently to be the main factor involved in fish motility activation among species [35, 36]. In a few fish species, osmolarity is acting in synergy with another factor such as specific ions [35]. In some marine species such as herring, activation of spermatozoa requires egg-derived substances. Two types of sperm-activating factors have been identified in Pacific herring, *Clupea pallasii*, eggs: a water-soluble protein released into the surrounding water [37] and a water-insoluble sperm motility-initiating factor localized in the vicinity of the micropylar opening of eggs [38].

3.3 The perception of the signal by the sperm membrane

Freshwater fish sperm cells when released into the surrounding water can increase their cytoplasmic volume in response to osmotic stress. In case of carp spermatozoa, the cell volume increases several times as a result of the influx of water [39]. Results of Cabrita et al. [40] show that small change of cell volume occurs in response to hypo-osmotic shock. The comparative study of Bondarenko et al. [41] puts forward a large species specificity regarding the osmotic reaction (swelling) among freshwater species and demonstrates that there is no change of trout sperm volume measured during the motility period. Altogether, the volume change, if any, represents a long-term osmotic reaction rather than the immediate signal for motility activation given its time delay (several tens of seconds) relative to the briefness of motility appearance (less than a second according to Prokopchuk et al. [33].

A series of experiments by Takei et al. [42] aims to better explore the respective roles of K^+ ions and osmolarity; this paper proposes a mechanism involving aquaporins and volume changes as a response to osmolarity stress. However, the volume changes measured by the authors are of low amplitude: the engendered volume difference is so low that it cannot be responsible of a physiological role in motility control. Furthermore, the time scale and the volume change measurements obtained at 5 min after motility activation correspond to time scales that are not fit with the previously published results of Prokopchuk et al. [33] showing that the motility response occurs about 100 ms after reception of the activation signal of fish spermatozoa. In addition, previous results by Bondarenko et al. [41] demonstrate that no significant volume change follows the motility activation of trout spermatozoa, a situation that contrasts with that of carp.

In many marine teleost species, hypertonicity induces the motility of spermatozoa. Nevertheless, an increase in external osmolality is sometimes not the only condition for motility activation of marine fish spermatozoa: in case of herring, this activation also needs the contact of sperm with egg-derived substances, facilitating fertilization [37, 38, 43, 44]. Sperm-activating factors are two types of in the Pacific herring, *Clupea pallasii*: a water-soluble herring egg protein [37, 43, 44] and water-insoluble initiating factor, from the vicinity of the micropylar opening of the egg [38]. In another group of marine fish species collectively named flatfishes, the main signal perceived by the sperm membrane is the CO_2 concentration [34] that is high in the genital tract but very low in sea water. The intracellular equilibrium between CO_2 and bicarbonate constitutes the second step in the control of intracellular ionic concentration that leads to regulation of flagellar motility [33].

3.4 The transduction of the signal across the membrane and the cytoplasm

In fish spermatozoa, external signals triggering sperm motility activation are acting at the level of spermatozoon plasma membrane, hyperpolarization/depolarization of membrane, and ion channels or aquaporins activity, but this topic is still challenging because of the scarcity of experimental results in this area. As mentioned above, it is clear that water transport itself is not a main process involved in fish sperm motility activation. Nevertheless, in the presence of aquaporins in the head and flagella plasma membrane of the seawater fish, gilthead sea bream (*Sparus aurata*) spermatozoa and their involvement in cAMP-mediated phosphorylation of axonemal proteins were established [45, 46]. In this species, the water efflux via aquaporins would determine a reduction in the cell volume, which would raise the intracellular concentration of ions. This would lead to the activation of adenylyl cyclase and motility initiation by cyclic AMP-dependent protein phosphorylation and dephosphorylation [46]. Such cascade of events remains hypothetical because of the timing of such process compared to the extreme briefness (less than 0.1 s) of the reaction of the axoneme activation [33].

The presence of different ion channels was described in sperm plasma membrane [34]. Cytosolic pH could be considered as another participant of signaling pathways, as it is known to be one of the parameters influencing sperm motility [47, 48]. Environmental conditions that inhibit spermatozoa motility can decrease the intracellular pH, resulting in a more acidic cytoplasm in nonmotile spermatozoa than in motile spermatozoa [49]. The decrease of internal pH in sperm would directly affect flagellar movement through inhibition of dynein activity. The involvement of the Na+/H+ exchangers in sperm motility activation process was reported for *Cyprinus carpio* [50] and was proposed for *Brycon henni* [51]. The former authors suggest that the regulation of the exchangers depends on osmolality conditions [52]. According to Krasznai et al. [53], an opening and closing of K^+ channels in the plasma membrane of the spermatozoon under hypo-osmosis-induced initiation of sperm motility is resulting in a remarkable local hyperpolarization or depolarization of the spermatozoon plasma membrane. Such transient depolarization may open Ca^{2+} channels, resulting in an influx of Ca^{2+} and activation of the flagellar motility of carp sperm.

The involvement of K^+ and Ca^{2+} transport through ion channels at the plasma membrane of spermatozoa in the triggering of the motility initiation has also been shown for rainbow trout (*Salmo gairdneri*) spermatozoa [54]. As for Na^+ channels, Tanimoto and Morisawa [54] supposed that Na^+ channels do not play an important role in sperm motility in rainbow trout, although they did not exclude its possible involvement in sperm motility control, for example, through Na^+-H^+ exchange.

Altogether, the precise mechanisms of regulation of ion channel activity and their participation in the hyperpolarization of the spermatozoon membrane, which is associated with the activation of sperm motility [35, 53], remain poorly understood. What are the next-step processes occurring at the level of membrane leading the subsequent activation of the axoneme?

3.5 Transduction and reception of the signal at the axoneme level

Among other processes, an increase of intracellular concentration of ions could lead to the activation of adenylyl cyclase, which in turn would determine the motility initiation by a cAMP-dependent protein phosphorylation and dephosphorylation mechanism [46]. It is known that, in mammals, protein tyrosine phosphorylation of several proteins is upregulated by reactive oxygen species (ROS). ROS (especially H_2O_2) may enhance tyrosine phosphorylation through the selective suppression of tyrosine phosphatase activity [55] or activation of adenylyl cyclase, thus producing a higher cAMP level and leading to the subsequent activation of the serine/threonine kinase A [56].

Cyclic AMP is an important factor in the activation process of fish spermatozoa. The link between cAMP concentration increase and motility initiation at the axoneme level was mainly investigated in Salmonidae. It involves a complex series of phosphorylation and dephosphorylation events. This includes the cAMP-dependent phosphorylation of the 15 kDa movement-initiating phosphoprotein [57, 58] of a PKA [59] and of the 22 kDa dynein light chain [60]. Protein phosphorylation is also regulated by proteasomes [60, 61]. The precise dependence between protein phosphorylation and microtubule sliding and movement initiation is still under investigation. A Ca^{2+}-mediated and/or cAMP-dependent phosphorylation signaling mechanism through the radial spoke/central pair system of the axoneme has been proposed [62, 63].

The potential role of reactive oxygen species (ROS) generated at the contact of sperm with aerobic condition such as the external medium at ejaculation and the molecular mechanisms by which these reactive metabolites exert their biological activity has been put forward by Baker and Aitken [64]. A gas, NO, was also observed to enhance motility of fathead minnow spermatozoa [65]. Nevertheless, the mechanism by which NO affects sperm motility is probably unrelated to osmolality, because of its very low active concentration range.

3.6 The motility period

During the period after spermatozoa comes in contact with an activating medium, the ion concentrations inside the sperm cell are rebalanced, and osmotic pressure affecting the sperm membrane becomes harmful for sperm integrity, limiting the period of motility to a short interval [32]. These phenomena are much faster and more obvious in freshwater species, in which sperm motility usually does not last for more than 0.5–2 min [32].

Video 1 can be viewed at https://vimeo.com/310085381.

Brook trout (*Salvelinus fontinalis*) spermatozoa recorded while swimming in a swimming medium composed of low osmolarity (10 mM Tris at pH 8.0 + 10 mM $CaCl_2$). Remark the briefness of motility (around 30 s duration at 10°C) (courtesy of Dr. Galina Prokopchuk).

In case of sperm collection for artificial reproduction or in vitro studies, particular care should be taken to avoid precocious activation of motility: one should avoid any contact of sperm cells with external water [66] or urine during stripping [67, 68]. Any assessment of motility parameters should be started as soon as possible after sperm activation, bearing in mind that the earliest period of motility (the most efficient one) should be characterized. Fish spermatozoa are usually characterized by a very high initial velocity (up to 200 μm s^{-1}) due to the high flagellar beat frequency (up to 100 Hz; [32]); this "most active" period lasts only a few seconds immediately after contact with the activating medium.

Values of all motility parameters decrease rapidly immediately after initiation of fish sperm flagella movement [36, 69], which is why any motility parameter must refer to a precise time point after activation for intra- or interspecies comparisons [36, 69, 70]. Generally, the duration of motility is a trade-off between the level of energy stocks possessed by a cell and the process of osmotic damage experienced by this cell. The latter is more critical in freshwater fish species, and the former is important for marine fish [36]. At a precise time point, most spermatozoa have very similar characteristics [69, 70].

Video 2 can be viewed at: https://vimeo.com/310086859.

Carp (*Cyprinus carpio*) spermatozoa recorded while swimming in two different media successively; the first part of the record shows the carp sperm population activated in a swimming solution composed of 45 mM NaCl +5 mM KCl + Tris at pH 8.0, while the second part shows spermatozoon activation in a low-osmolarity medium (distilled water + Tris at pH 8.0). Remark the curling of the flagellar tip which limits the duration of motility at low osmolarity (courtesy of Dr. Volodymyr Bondarenko).

4. Energetic of sperm motility

Storage of energy mostly results from mitochondrial respiration that generates ATP. Energy metabolism also involves other compounds such as creatine phosphate that contributes to the maintenance of the intracellular energy level in connection with ATP.

4.1 Mitochondrial respiration

In many fish species, measurement of respiratory activity presents difficulties because of the low oxygen consumption of spermatozoa, in contrast to model species such as sea urchin [71]; in addition, the low respiratory activity remains almost unchanged when fish spermatozoa are transferred into motility-activating solutions [72], while it is about 50-fold increased when sea urchin sperm is transferred into sea water [71]. Efficient respiration needs to be coupled in mitochondria to ATP production via the ATP synthase [73]. For estimation of the full respiratory capacity of mitochondria, it is useful to apply diffusible "uncouplers" such as CCCP or FCCP (carbonylcyanide-4-trifluromethoxy-phenylhydrazone): these compounds are diffusible through the membranes and allow full rate of electron transfer in the electron chain of mitochondria without restriction due to its control by ATP synthase. The effects of respiratory inhibitors such as oligomycin [73] or KCN and their relationship with ATP stores of fish sperm were studied in details by Dreanno et al. [74]. Mitochondrial inhibitors have little effect in case of trout [75–78] or turbot [79] spermatozoa. Respiration rate in quiescent fish spermatozoa (before motility activation) needs to be only minimal but enough to maintain this ATP level prior to ejaculation. Such low but substantial respiration is enough for basal metabolism to maintain ionic exchanges and balances across the plasma membrane [8, 76, 79].

4.2 Generation and storage of ATP

Generation of ATP by mitochondria occurs by electron transfer along the mitochondrial respiratory chain that generates a proton gradient across the inner mitochondrial membrane. Dissipation of the proton gradient occurs by passage of H^+ through a specific ATP synthase localized at the inner part of the mitochondrial membrane [8, 73]. ATP thus accumulated is transported out of the mitochondrion by a translocase (ATP-ADP exchanger) toward the flagellar compartment; then, ATP molecules diffuse along the axoneme possibly assisted by a carrier device called energy shuttle as explained in Section 4.4.

The energy stored in fish sperm prior to and used during the motility period was evaluated in several fish species [80–82]. This was described in turbot, for example [79, 83, 84], sea bass [85] perch [86], bluegill [87, 88], trout [71, 89, 90], carp [39, 91], sturgeon [92, 93], and catfish [94]. Values for spermatozoa of other fish species can also be found in Cosson [8, 32, 95], Dzyuba et al. [30], and Ingermann [96].

4.3 Consumption of ATP during the motility period

ATP is the only high-energy compound that is hydrolyzed by the motor protein of the axoneme called dynein ATPase [16, 17, 97]. Dyneins are macromolecular assemblies of more than 1 MDa molecular weight linearly bound to each microtubule doublet of the axoneme and which function is mechanochemical. Its role it to collect the chemical energy generated by hydrolysis of an ATP molecule so as to induce a trans-conformation that constitutes the elementary step of movement generation. Rows of dyneins positioned along each microtubule amplify in a cooperative way this elementary movement so as to provoke the sliding between adjacent microtubules of the axoneme. Differential sliding generates bending, and the bending waves that propagate along the flagellum (usually from the head to the tip of the axoneme) result in a translational movement of the whole sperm cell [21].

In case of fish spermatozoa, dynein molecules have an intense activity: this fast activity results in a beat frequency (up to 80–100 Hz) much higher than in species other than fish. As a result, ATP is hydrolyzed at high speed, and the ATP store is rapidly decreasing, becoming partly exhausted at the end of the motility period [32].

4.4 Other energetic molecules assisting ATP maintenance

Following its synthesis by mitochondria, ATP should be transported and distributed all along the flagellum so as to supply chemical energy to sustain the mechanical energy generated by the dynein motors that are distributed all around the flagellar axoneme. Theoretical considerations lead to postulate the presence of a distribution system that ensures a constant ATP concentration at any point along the axoneme [98, 99]. Such shuttle system involves an additional high-energy compound and assistance of enzymatic system that was shown to be present in fish sperm cells [90] and is detailed below.

Even though the ATP molecule is the most common high-energy compound used as a fuel for many cell functions, including motility [8], several other high-energy molecules are present in living cells such as creatine phosphate or arginine phosphate. In fish, creatine phosphate (CrP) is the main high-energy compound that was characterized in spermatozoa of several fish species as a complement of ATP [74, 79, 85]. In the last decades, many studies have demonstrated the decrease of the ATP concentration inside the fish sperm cells during the motility period [39, 67, 74, 76, 79, 85, 90, 91, 93, 100]. A more restricted number of studies have investigated the concentration of ATP related compound such as ADP, AMP, CrP, and others [74, 79, 85, 100]. All these compounds are part of an intracellular network under control of different enzymes that are able to transfer high-energy phosphate bonds from one to another (see Figure XX in Chapter 1 of Cosson [21]), for example, the equilibrium ATP< = >ADP + Pi is catalyzed by enzymes called ATPases. In another example, ADP + CrP < = >ATP + Cr is controlled by enzymes called creatine kinases. One creatine kinase is mitochondrial, while a second one is in the flagellum and distributed all along the axoneme (**Figure 1**). The mitochondrial creatine kinase delivers CrP that diffuses along the flagellum, both creatine kinases being present in trout sperm cells. The rate of diffusion of CrP molecules is higher than that of ATP [99]. Such an arrangement of catalytic activities and substrates

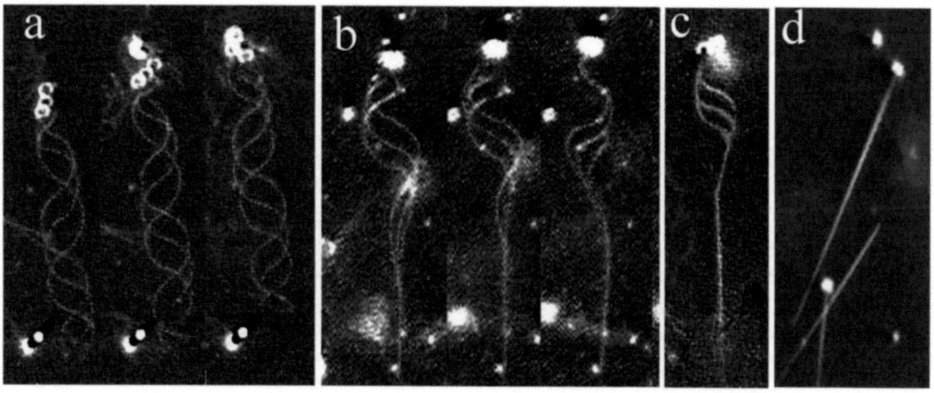

Figure 2.
Evolution of the flagellar shape of fish spermatozoa during the motility period. From left to right, successive positions of a turbot (Scophthalmus maximus) spermatozoon video recorded at different time points post-activation: (a) right after transfer in sea water, (b) 2 min later, (c) more than 3 min later, (d) after full stop. Dark field microscopy with stroboscopic illumination (150 Hz); 100× objective lens [32].

constitutes an intracellular network ensuring the correct production and distribution of energy in fish sperm cells (**Figure 1**) and is called the "ATP shuttle" [99].

A local axonemal paralysis appears in sperm flagella of many species during the motility period: this is an indication that the renewal of ATP is not fully efficient, and the production of ATP from CrP in this distal portion of the flagellum is too slow, while the proximal portion, that is close to the mitochondrion, can use directly the ATP still produced by the latter. Stiffening of the distal flagellar tip was observed in vivo (but not in reactivated sperm) in two sea urchin species [101] but also in trout sperm (Cosson, unpublished) after application of thiourea, an inhibitor of respiratory phosphorylation (**Figure 2**).

5. Control of fish sperm physiology by other factors

Changes in the environment of fish spermatozoa impose other chemicals as well as physical constrains: temperature is controlling sperm physiology at many levels such as membrane permeation, enzymatic activities, or energetic metabolism [102]. Fish species are exposed to a large variety of temperature conditions, especially during their reproduction period; thus, an optimal choice of temperature is a key parameter for controlling the best conditions for best adapted sperm physiology.

Viscosity of the swimming medium is also an important factor influencing the physiology of fish spermatozoa. According to physical laws, especially microfluidics, progression of the sperm cell occurs because of the friction of the flagellum against the external milieu, water being a quite viscous fluid. Higher viscosity is encountered by fish spermatozoa when getting close to the fluids surrounding the egg, which modifies and thus controls the swimming physiology of the sperm cells [21, 103] including female cryptic choice of sperm cells by the egg [104].

Among other factors that play a signaling role for fish physiology are some molecules that control the sperm/egg interplay at fertilization; these molecules are commonly called chemoattractants and are able to finely tune the motility function of fish spermatozoa so as to optimally guide sperm cells to egg which ultimately results in an increase of fertility success [21]. In case of fish, the spermatozoon must localize the entrance point at the surface of the egg so as to penetrate and deliver the male genome [105]. Ultimately, the male genome will combine with that of the female so as to constitute the zygote.

6. An update of modern technologies allowing fish sperm quality assessment

6.1 CASA systems

Evaluation of fish sperm swimming performances needs at first good quality video records so as to measure the distance covered by each sperm head for a time period corresponding, for example, to the time separating two successive video frames, such as 20 ms for the European video standard (**Figure 3**).

Several works in the last decade have reported the use of CASA systems to assess sperm motility in fish [106]. Based on the integration by the computer of successive positions of the moving head of spermatozoa in consecutive frames of video records to calculate the trajectories and their characteristics, CASA describes different parameters of sperm swimming linked to velocity, for example, VCL or instant speed (frame to frame displacement) along the real track, VAP or velocity along a smoothed track, VSL or progressive velocity following the straight line from the origin to the end of the track during the corresponding period of time and other parameters linked to the wobble of sperm head such as Mean angular displacement (MAD), amplitude of lateral head displacement (ALH) and beat cross frequency (BCF) and linearity. Lastly, the ratio between average path and straight-line path is used to describe the straightness of the trajectories (reviewed in Rurangwa et al. [107]).

The CASA system recently developed by Wilson-Leedy and Ingermann [108] as a plug-in to image J software freely available from NIH site (http://rsb.info.nih.gov/ij/plugins/casa.html) has been tested in different species including zebrafish [109]. More recently, it was improved and adapted to trout sperm motility by Purchase and Earle [110].

One important aspect for motility estimation of sperm quality by CASA is that the practical conditions employed to perform such tests are in several respects not reflecting the natural situation. Many broadcast spawners like salmon or trout

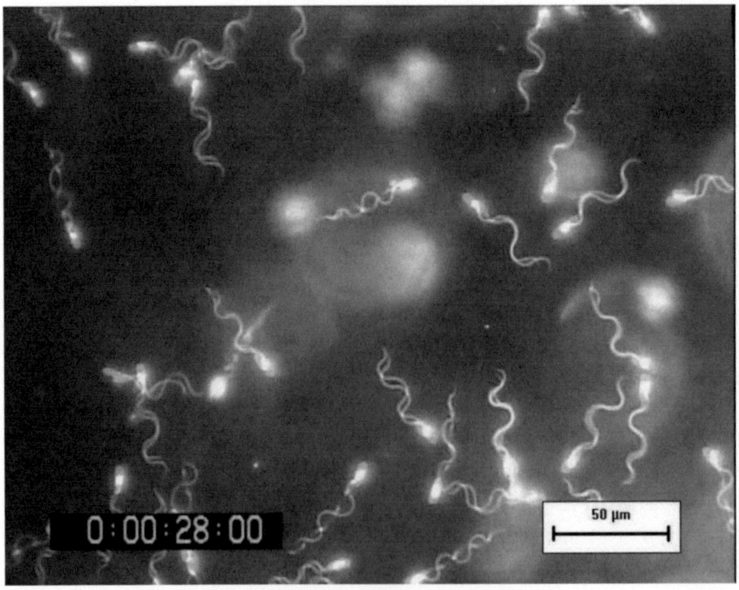

Figure 3.
Video image of swimming sturgeon spermatozoa. Dark-field video microscopy with stroboscopic illumination. Flashes are every 10 ms. At bottom right is the indication from the stop-watch giving the time (in milliseconds) spent since motility activation.

reproduce in highly turbulent water, which certainly influences the sperm/egg meeting chances and thus the fertilization success. Effect of such turbulent water shear at some optimal values was studied using biophysical methods by Crimaldi and Browning [111], and it was shown experimentally to increase the proportion of fertilized eggs in sea urchin [112].

Also, for practical reasons, CASA records are obtained in conditions where spermatozoa swim in the vicinity of glass surfaces: such situation was shown in literature to affect motility parameters [113, 114]. An important consequence of this is that the motility parameters mostly refer to a situation where sperm cells swim in a planar manner; it is well known since early studies [115] that when spermatozoa swim freely far from any surface, they adopt a helical trajectory. The latter was shown to decrease up to 10-fold the efficient gross velocity but surprisingly to increase around 6-fold the fertilization kinetics [116]. All these findings emphasize several limits of the application of these CASA systems when used to predict quality of sperm regarding the fertilization rate.

The comparative values of sperm velocity among fish species, including salmonid, can be found in Cosson [32] with respect to motility duration and ATP stores prior to activation.

6.2 Evaluation of flagella performances

Behavior of the flagellum determines the motility guideline of the spermatozoon, so the description of intrinsic flagellar wave properties is considered as one of the most informative methods for assessing and controlling sperm motility. In order to observe the detailed pattern of live flagella or of their major components, it was proposed to use phase contrast or dark field optical microscopy with high magnification (40×–100×) objective lenses, which, if applied with oil immersion, result in a bright image of the very small diameter object that constitutes a flagellum. To achieve complementary assessment, additional methods, such as stroboscopic illumination or high-speed video techniques, allow to record sperm during its motion and specially to obtain flagellar images of high quality and resolution. Multiflash stroboscopic illumination thus allows visualization on each frame of well-defined successive positions of a same moving spermatozoon at time intervals ranging in milliseconds. Alternatively, high-speed video recording provides higher spatial and temporal resolutions (up to several 1000 images/s). Serial frames individually selected from such video records allow to follow successive positions (every millisecond or less) of flagellum waves covering one or several full beat cycles [32].

Video 3 can be viewed at https://vimeo.com/309942086.

Legend of the video: a sturgeon spermatozoon was recorded by phase-contrast video microscopy according to conditions similar to those described in **Figure 4** (courtesy of Dr. Bondarenko Volodymyr). Spermatozoa of Siberian sturgeon (*Acipenser baerii*) were activated by dilution in pond water and recorded 10 s after activation. The length of flagella is about 50 μm long. Image rate is 100-fold slower than real. Remark that some spermatozoa show abnormal shape.

Evaluation of sperm flagellum performances on a large variety of fish species leads to a series of predictions briefly summarized below (see **Figure 5**):

- The swimming velocity is linearly proportional to the flagellar beat frequency.
- The smaller the flagellum length, the lower the swimming velocity.
- The number of waves along the flagellum varies linearly with the flagellum length.

- The power output of the flagellum varies in proportion with the flagellar beat frequency.
- The flagellar length has no significant influence on the swimming velocity.
- The swimming efficiency increases in proportion with the swimming speed.
- The wave amplitude and the swimming velocity are linearly related.
- The swimming velocity is lowly dependent of the medium viscosity, but the latter greatly affects the wave shape.
- Waves of helical shape generate rotation of the sperm cell at a frequency that is in proportion with the flagellar beat frequency but less efficient than planar waves.

While the whole cell is moving from left to right, the flagellum wave progresses from right to left. The wavelength is defined as the distance "L" between two inflection points. The half wave amplitude is represented as "A" and the bend angle of each wave as "a". The number of waves generated every second is called the beat frequency. Such parameters are quantified and used to characterize fish sperm cells exposed to various swimming situations affecting their physiology.

Two other examples of high-speed video records of fish spermatozoa are presented below.

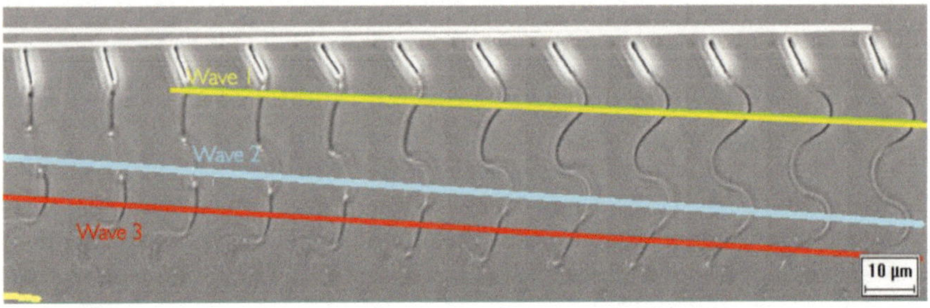

Figure 4.
Successive positions of a sturgeon sperm flagellum recorded by high-speed video microscopy. Initial record was at 5000 images/s with a 100× phase-contrast lens and an Olympus high-speed video camera. In this panel from left to right, successive images collected every millisecond are presented so as to show the wave propagation of three successive waves and the minor progression of the head tip (white straight lines) during this short time period.

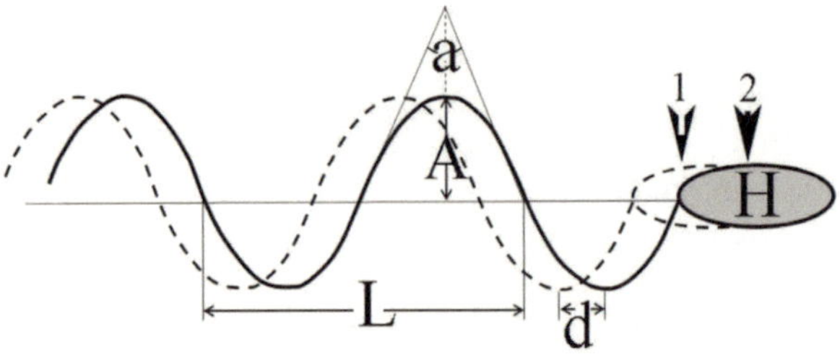

Figure 5.
Flagellar parameters of a swimming fish spermatozoon. The sperm cell, such as that recorded according to ***Figure 3****, is represented in two successive positions (1–2) separated by a short time period (0.5 ms, for example).*

Video 4 can be viewed at https://vimeo.com/310100596.

Pangasius (*Pangasianodon hypophthalmus*) spermatozoa recorded by high-speed video combined with dark-field microscopy and was visualized 10-fold slower than normal. Sperm movement was recorded at 13 s post-activation in a 10% sea water solution. Length of flagellum is about 50 μm (courtesy of Dr. Galina Prokopchuk).

Video 5 can be viewed at https://vimeo.com/310102833.

Tilapia (*Sarotherodon melanotheron heudelotii*) spermatozoon recorded by high-speed video microscopy and visualized 20-fold slower. Sperm movement was recorded at 18 s post-activation in 50% sea water containing bovine serum albumin (BSA at 0.5%) to prevent sticking to the glass slide. Notice the short sperm flagellum in this species and the swollen midpiece, mostly composed of mitochondria (courtesy of Dr. Galina Prokopchuk).

6.3 Biochemical methods

6.3.1 Respiration, oxidative stress, free radicals, and DNA damage

Fish spermatozoa present a low respiratory rate that makes this evaluation quite delicate because of the sensibility of detection methods. Respiratory activity commonly uses oxygen electrodes that measure the oxygen concentration of a small volume of sperm suspension. These media are either preventing or activating motility and can be complemented by different effectors (substrates, uncouplers, or inhibitors) of respiration. In general, mitochondrial inhibitors have little effect on the motility of fish spermatozoa. Sperm oxygen consumption rate published in literature was presented in a comparative way for about 10 different fish species by Ingermann [96] and shows that values present a large variability from 1.4 to 70 nmoles O_2/min/10^9 spermatozoa depending on species.

The contact of fish sperm with the external milieu occurring at ejaculation leads to exposure of sperm cells to high concentration of oxygen, provoking different kinds of stress [117]. Among other chemicals responsible of stress, reactive oxygen species are highly aggressive [118, 119]. The oxidative stress can be evaluated by several methods. Results of these studies show that several protections against the oxidative stress are present in fish sperm cells and in the seminal fluid [31, 120].

Also, the presence of free radicals leads to DNA damage during stressing situations such as those due to application of cryo-techniques that influence the quality of the progeny [121]. Recent studies on genes and protein expression of fish sperm cells show that both are controlled by various factors of the external milieu such as salinity or timing during the reproductive season [122, 123]. Among other factors, the level of phosphorylation of specific proteins constitutes important signaling factors controlling function efficiency of fish spermatozoa [124, 125]. Proteomics represent a promising approach to study specific physiological situations encountered by fish spermatozoa [126].

6.3.2 Evaluation of energetic compounds concentration

ATP content of sperm cells can be evaluated by several methods, including measurement in a single spermatozoon cell as recently shown by Chen et al. [127]. The most popular method for evaluation of the ATP content classically uses a coupling with the light-emitting system composed of luciferin and luciferase. A full evaluation of the storage of energy in fish sperm cells needs the determination not only of the internal content of ATP but also that of other energetic compounds that are able to exchange high-energy phosphate bonds able to be transferred to ADP and allow to reconstitute the intracellular ATP store. Such evaluation was established in case of sturgeon sperm [100, 128] by the use of liquid chromatography combined to HPRS or in case of turbot or sea bass sperm where the adenine nucleotides' energetic balance was determined by H^+-NMR and ^{31}P-NMR analysis, [74, 79] as well as in trout by ^{31}P-NMR [89, 90]. All these results clearly point out to the fact that ATP level can be rescued by the CrP generated by the mitochondrial metabolism. This means that other phosphagen compounds are as important as ATP in the energy balance of fish sperm cells [129].

7. Conclusion

Fish sperm physiology is under control of various parameters of the external milieu: the latter is subjected to changes due to the different environmental conditions that sperm cells have to deal with such as (1) the ionic concentration of internal as well as external fluids, (2) the pH, (3) the osmolarity, (4) the temperature, and (5) specific molecules acting as signals such as chemoattractants that control the sperm-egg interaction at fertilization [21]. In fish spermatozoa, the interplay between the different actors results in a complex signaling network that exquisitely optimizes the various functions of fish spermatozoa, especially that of motility, in a large variety of situations. A better understanding of this complex network is important so as to decrease the effects of possible damage (osmotic, oxidative, etc.) when fish sperm cells are exposed to drastic conditions such as those imposed during application of cryopreservation methods.

Acknowledgements

The author is thankful to the Faculty of Fisheries and Protection of Waters, Univ. of South Bohemia in Ceske Budejovice, Czech Republic, especially to Dr. Galina Prokopchuck and Dr. Volodymyr Bondarenko for their high-speed video records of fish spermatozoa. The study was financially supported by the Ministry of Education, Youth and Sports of the Czech Republic projects CENAKVA (LM2018099), by project Biodiversity (CZ.02.1.01./0.0/0.0/16_025/0007370 Reproductive and genetic procedures for preserving fish biodiversity and aquaculture), and by the Grant Agency of the University of South Bohemia in Ceske Budejovice (125/2016/Z).

Author details

Jacky Cosson
University of South Bohemia, Ceske Budejovice, Czech Republic

*Address all correspondence to: jacosson@gmail.com

IntechOpen

© 2019 The Author(s). Licensee IntechOpen. This chapter is distributed under the terms of the Creative Commons Attribution License (http://creativecommons.org/licenses/by/3.0), which permits unrestricted use, distribution, and reproduction in any medium, provided the original work is properly cited.

References

[1] Cosson J, Prokopchuk G, Inaba K. The flagellar mechanics of spermatozoa and its regulation. In: Cosson J, editor. Flagellar Mechanics and Sperm Guidance. Sharjah, UAE: Bentham Science Publisher Ltd; 2015. pp. 3-134

[2] Jamieson BGM. Fish evolution and systematics: Evidence from spermatozoa. In: With a Survey of Lophophorate, Echinoderm and Protochordate Sperm and an Account of Gamete Cryopreservation. Cambridge: Cambridge University Press; 1991. 319 p. ISBN: 0-521-41304-4

[3] Prokopchuk G, Cosson J. Biophysics of fish sperm flagellar movement: Up-dated knowledge and original directions. In: Jimenez-Lopez JC, editor. Cytoskeleton—Structure, Dynamics, Function and Disease. London, UK: IntechOpen Access Publisher; 2017. pp. 127-150. ISBN: 978-953-51-3170-0

[4] Bondarenko V, Prokopchuk G, Cosson J. Fish sperm flagella: Original features and biological implications through the lens of modern technologies. In: Uzbekov R, editor. Flagella and Cilia: Types, Structure and Functions. New York, USA: Nova Publisher; 2018. pp. 49-82

[5] Linhartova Z, Rodina M, Nebesarova J, Cosson J, Psenicka M. Morphology and ultrastructure of beluga (*Huso huso*) spermatozoa and a comparison with related sturgeons. Animal Reproduction Science. 2013;**137**:220-229. DOI: 10.1016/j.anireprosci.2013.01.003

[6] Gibbons BH, Gibbons IR, Baccetti B. Structure and motility of the 9 + 0 flagellum of eel spermatozoa. Journal of Submicroscopic Cytology. 1983;**15**:15-20

[7] Gibbons BH, Baccetti B, Gibbons IR. Live and reactivated motility in the 9 + 0 flagellum of anguilla sperm. Cell Motility. 1985;**5**:333-350. DOI: 10.1002/cm.970050406

[8] Cosson J. Fish spermatozoa motility: Physical, and bio-energetic interactions with their surrounding media. In: Morisawa M, editor. Sperm Cell Research in the 21st Century: Historical Discoveries to New Horizons. Tokyo, Japan: Adthree Publisher Co.; 2012. pp. 152-156. ISBN: 978-4-9044419-37-3

[9] Billard R. Ultrastructure of trout spermatozoa: Changes after dilution and deep-freezing. Cell and Tissue Research. 1983;**228**:205-218

[10] Shiba K, Mizuno K, Inaba K. Molecular composition of the axonemal components between sperm flagella and *Chlamydomonas* flagella. In: Morisawa M, editor. Sperm Cell Research in the 21st Century: Historical Discoveries to New Horizons. Tokyo, Japan: Adthree Publisher Co.; 2012. pp. 30-40. ISBN 978-4-9044419-37-3

[11] Inaba K. Molecular architecture of the sperm flagella: Molecules for motility and signaling. Zoological Science. 2003;**20**:1043-1056. DOI: 10.2108/zsj.20.1043

[12] Cosson J. A moving image of flagella: News and views on the mechanisms involved in axonemal beating. Cell Biology International. 1996;**20**:83-94

[13] Mattei C, Mattei X. Spermiogenesis and spermatozoa of the Elopomorpha (teleost fish). In: Afzelius B, editor. The Functional Anatomy of the Spermatozoon. Oxford: Pergamon Press; 1975. pp. 211-221

[14] Smith EF, Yang P. The radial spokes and central apparatus: Mechano-chemical transducers that regulate flagellar motility. Cell Motility and the Cytoskeleton. 2004;**57**:8-17. DOI: 10.1002/cm.10155

[15] Heuser T, Raytchev M, Krell J, Porter ME, Nicastro D. The dynein regulatory complex is the nexin link and a major

regulatory node in cilia and flagella. The Journal of Cell Biology. 2009;**187**: 921-933. DOI: 10.1083/jcb.200908067

[16] Gibbons IR, Rowe AJ. Dynein: A protein with adenosine triphosphatase activity from cilia. Science. 1965;**149**: 424-426. DOI: 10.1126/science.149.3682.424

[17] Gibbons IR. Discovery of dynein and its properties: A personal account. In: King SM, editor. Cambridge, MA, USA: Dyneins published by Academic Press; 2017. pp. 629-639. DOI: 10.1016/B978-0-12 809471-6.00001-2. https://doi.org/10.1016/B978-0-12-809471-6.00001-2

[18] Dillon RH, Fauci LJ. An integrative model of internal axoneme mechanics and external fluid dynamics in ciliary beating. Journal of Theoretical Biology. 2000;**207**:415-430. DOI: 10.1006/jtbi.2000.2182

[19] Gibbons BH, Gibbons IR. Calcium induced quiescence in reactivated sea urchin sperm. Journal of Cell Biology. 1980;**84**:13-27. DOI: 0021-9525/80/01/0013/15

[20] Eshel D, Brokaw CJ. New evidence for a "biased baseline" mechanism for calcium-regulated asymmetry of flagellar bending. Cell Motility and the Cytoskeleton. 1987;**7**:160-168. DOI: 10.1002/cm.970070208

[21] Cosson J, editor. Flagellar Mechanics and Sperm Guidance. Charjah, UAE: Bentham Books Publisher; 2015. 424 p. DOI: 10.2174/97816810812811150101

[22] Sale WS, Satir P. Direction of active sliding of microtubules in *Tetrahymena cilia*. Proceedings of the National Academy of Sciences of the United States of America. 1977;**74**(5):2045-2049. DOI: 10.1073/pnas.74.5.2045

[23] Billard R. Spermatogenesis and spermatology of some teleost fish species. Reproduction Nutrition Development. 1986;**26**(4):877-920

[24] Billard R. La spermatogenese de *Poecilia reticulata* I. Estimation du nombre de generations goniales et rendement de la spermatogenese. Annales de Biologie Animale Biochimie Biophysique. 1969;**9**:251-271

[25] Schulz RW, Renato de França L, Lareyre JJ, Le Gac F, Chiarini-Garcia H, Nobrega RH, et al. Spermatogenesis in fish. General and Comparative Endocrinology. 2010;**165**:390-411

[26] Kozminski KG, Johnson KA, Forscher P, Rosenbaum JL. A motility in the eukaryotic flagellum unrelated to flagellar beating. Proceedings of the National Academy of Sciences of the United States of America. 1993;**90**:5519-5523

[27] Morisawa S, Morisawa M. Induction of potential for sperm motility by bicarbonate and pH in rainbow trout and chum salmon. Journal of Experimental Biology. 1988;**136**:13-22

[28] Morisawa S, Ishida K, Okuno M, Morisawa M. Role of pH and cyclic adenosine monophosphate in the acquisition of potential for sperm motility during migration from the sea to the river in chum salmon. Molecular Reproduction and Development. 1993;**34**:420-426

[29] Redondo C, Cosson MP, Cosson J, Billard R. *In vitro* maturation of the potential for movement of carp spermatozoa. Molecular Reproduction and Development. 1991;**29**:259-270

[30] Dzyuba B, Cosson J, Dzyuba V, Fedorov P, Bondarenko O, Rodina M, et al. Sperm maturation in sturgeon (Actinopterygii, Acipenseriformes): A review. Theriogenology. 2017;**97**:134-138

[31] Dzyuba B, Cosson J, Boryshpolets S, Bondarenko O, Prokopchuk G, Gazo I, et al. *In vitro* maturation of spermatozoa in sterlet *Acipenser ruthenus*. Reproductive Biology. 2014;**14**(2):160-163

[32] Cosson J. Frenetic activation of fish spermatozoa flagella entails short-term motility, portending their precocious decadence. Journal of Fish Biology. 2010;**76**:240-279. DOI: 10.1111/J.1095-8649.2009.02504

[33] Prokopchuk G, Dzuba B, Boryshpolet S, Linhart O, Cosson J. Motility initiation in fish spermatozoa: Description of the propagation of very first initial waves. Theriogenology. 2015;**84**(1):51-61. DOI: 10.1016/j.theriogenology.2015.02.011

[34] Inaba K, Dreano C, Cosson J. Control of sperm motility by CO_2 and carbonic anhydrase in flatfish. Cell Motility and the Cytoskeleton. 2003;**55**:174-187

[35] Alavi SMH, Cosson J. Sperm motility in fishes (II) effects of ions and osmolality: A review. Cell Biology International. 2006;**30**(1):1-14

[36] Cosson J. The ionic and osmotic factors controlling motility of fish spermatozoa. Aquaculture International. 2004;**12**(1):69-85

[37] Morisawa M, Tanimoto S, Ohtake H. Characterization and partial purification of sperm-activating substances from eggs of herring, *Clupea pallasi*. Journal of Experimental Zoology. 1992;**264**(2):225-230

[38] Yanagimachi R, Cherr GN, Pillai MC, Baldwin JD. Factors controlling sperm entry into the micropyle of salmonid and herring eggs. Development, Growth and Differentiation. 1992;**34**(4):447-461

[39] Perchec-Poupard G, Gatti JL, Cosson J, Jeulin C, Fierville F, Billard R. Effects of extracellular environment on the osmotic signal transduction involved in activation of motility of carp spermatozoa. Journal of Reproduction and Fertility. 1997;**110**:315-327

[40] Cabrita E, Alvarez R, Anel E, Herraez MP. The hypoosmotic swelling test performed with coulter counter: A method to assay functional integrity of sperm membrane in rainbow trout. Animal Reproduction Science. 1999;**55**:279-287

[41] Bondarenko O, Dzyuba B, Cosson J, Yamaner G, Prokopchuk G, Psenicka M, et al. Volume changes during the motility period of fish spermatozoa: Inter-specific differences. Theriogenology. 2013;**79**:872-881

[42] Takei GL, Mukai C, Okuno M. Regulation of salmonid fish sperm motility by osmotic shock-induced water influx across the plasma membrane. Comparative Biochemistry and Physiology. 2015;**182**:84-92

[43] Oda S, Igarashi Y, Manaka K, Koibuchi N, Sakai-Sawada M, Sakai K, et al. Sperm-activating proteins obtained from the herring eggs are homologous to trypsin inhibitors and synthesized in follicle cells. Developmental Biology. 1998;**204**:55-63

[44] Yoshida K, Inaba K, Ohtake H, Morisawa M. Purification and characterization of prolyl endopeptidase from pacific herring, *Clupea pallasi*, and its role in the activation of sperm motility. Development, Growth and Differentiation. 1999;**41**:217-225

[45] Zilli L, Schiavone R, Chauvigné F, Cerdà J, Storelli C, Vilella S. Evidence for the involvement of aquaporins in sperm motility activation of the teleost gilthead sea bream (*Sparus aurata*). Biology of Reproduction. 2009;**81**:880-888

[46] Zilli L, Beirão J, Schiavone R, Herraez MP, Cabrita E, Storelli C, et al. Aquaporin inhibition changes protein phosphorylation pattern following sperm motility activation in fish. Theriogenology. 2011;**76**:737-744

[47] Gatti JL, Christen R. Regulation of internal pH of sea urchin sperm. A role for Na/K pump. The Journal of Biological Chemistry. 1985;**260**:7599-7602

[48] Christen R, Schackmann RW, Shapiro BM. Ionic regulation of sea urchin sperm motility, metabolism and fertilizing capacity. Journal of Physiology. 1986;**379**:347-365

[49] Garcia MA, Meizel S. Regulation of intracellular pH in capacitated human spermatozoa by a Na^+/H^+ exchanger. Molecular Reproduction and Development. 1999;**52**:189-195

[50] Márián T, Krasznai Z, Balkay L, Balazs M, Emri M, Trón L. Role of extracellular and intracellular pH in carp sperm motility and modifications by hyperosmosis or regulation of the Na^+/H^+ exchanger. Cytometry. 1997;**27**:374-382

[51] Tabares J, Ruíz T, Arboleda L, Olivera M. Effect of some ions on sperm activation in *Brycon henni* (Eigenmann 1913). Acta Biológica Colombiana. 2007;**12**:87-98

[52] Marian T, Krasznai Z, Balkay L, Emri M, Trön L. Role of extracellular and intracellular pH in carp sperm motility and modifications by hyperosmosis of regulation of the Na^+/H^+ exchanger. Cytometry. 1997;**27**:374-382

[53] Krasznai Z, Márián T, Izumi H, Damjanovich S, Balkay L, Trón L, et al. Membrane hyperpolarization removes inactivation of Ca^{2+} channels, leading to Ca^{2+} influx and subsequent initiation of sperm motility in the common carp. Proceedings of the National Academy of Sciences of the United States of America. 2000;**97**:2052-2057

[54] Tanimoto S, Morisawa M. Roles for potassium and calcium channels in the initiation of sperm motility in rainbow trout. Development, Growth and Differentiation. 1988;**30**:117-124

[55] Aitken RJ, Bennets L. Reactive oxygen species: Friend or foe. In: De Jonge CJ, Barratt CLR, editors. The Sperm Cell–Production, Maturation, Fertilization, Regeneration. New York: Cambridge University Press; 2006. pp. 170-193

[56] Aitken RJ. Molecular mechanisms regulating human sperm function. Molecular Human Reproduction. 1997;**3**:169-173

[57] Hayashi H, Yamamoto K, Yonekawa H, Morisawa M. Involvement of tyrosine protein kinase in the initiation of flagellar movement in rainbow trout spermatozoa. The Journal of Biological Chemistry. 1987;**262**:16692-16698

[58] Jin ZX, Nakajima T, Morisawa M, Hayashi H. Isolation and properties of a protein complex containing flagellar movement-initiating phosphoprotein from testes of a white salmon. The Journal of Biochemistry. 1994;**115**:552-556

[59] Itoh A, Inaba K, Ohtake H, Fujinoki M, Morisawa M. Characterization of a cAMP-dependent protein kinase catalytic subunit from rainbow trout spermatozoa. Biochemical and Biophysical Research Communications. 2003;**305**:855-861

[60] Inaba K, Morisawa S, Morisawa M. Proteasomes regulate the motility of salmonid fish sperm through modulation of cAMP-dependent phosphorylation of an outer arm dynein light chain. Journal of Cell Science. 1998;**111**:1105-1115

[61] Inaba K, Akazome Y, Morisawa M. Purification of proteasomes from salmonid fish sperm and their localization along sperm flagella. Journal of Cell Science. 1993;**104**:907-915

[62] Harrison A, Sakato M, Tedford HW, Benashski SE, Patel-King RS, King SM. Redox-based control of the γ heavy chain ATPase from chlamydomonas outer arm dynein. Cell Motility and the Cytoskeleton. 2002;**52**:131-143

[63] Inaba K. Sperm flagella: Comparative and phylogenetic perspectives of protein components. Molecular Human Reproduction. 2011;**17**:524-538

[64] Baker MA, Aitken RJ. The importance of redox regulated pathways in sperm cell biology. Molecular and Cellular Endocrinology. 2004;**216**:47-54

[65] Creech MM, Arnold EV, Boyle B, Muzinich MC, Montville C, Bohle DS, et al. Sperm motility enhancement by nitric oxide produced by the oocytes of fathead minnows, *Pimephales promelas*. Journal of Andrology. 1998;**19**:667-674

[66] Perchec G, Cosson MP, Cosson J, Jeulin C, Billard R. Morphological and kinetic changes of carp (*Cyprinus carpio*) spermatozoa after initiation of motility in distilled water. Cell Motility and the Cytoskeleton. 1996;**35**:113-120. DOI: 10.1002/(SICI)1097-0169

[67] Perchec G, Cosson J, Andre F, Billard R. Degradation of the quality of carp sperm by urine contamination during stripping. Aquaculture. 1995;**129**:135-136. DOI: 10.1016/0044-8486(95)91958-X

[68] Dreanno C, Suquet M, Desbruye'res E, Cosson J, Le Delliou H, Billard R. Effect of urine on semen quality in turbot (*Psetta maxima*). Aquaculture. 1998;**169**:247-262. DOI: 10.1016/S0044-8486(98)00262-2

[69] Cosson J, Billard R, Gibert C, Dreanno C, Suquet M. Ionic factors regulating the motility of fish sperm. In: Gagnon C, editor. The Male Gamete: From Basic to Clinical Applications. Vienna, Illinois: Cache River Press; 1999. pp. 161-186

[70] Cosson J, Billard R, Cibert C, Dreanno C, Linhart O, Suquet M. Movements of fish sperm flagella studied by high speed videomicroscopy coupled to computer assisted image analysis. Polish Archives of Hydrobiology. 1997;**44**:103-113

[71] Christen R, Schackmann RW, Shapiro BM. Metabolism of sea urchin sperm. Interrelationships between intracellular pH, ATPase activity, and mitochondrial respiration. The Journal of Biological Chemistry. 1983;**258**(9):5392-5399

[72] Ulloa-Rodríguez P, Figueroa E, Díaz R, Lee-Estevez M, Short S, Farías JG. Mitochondria in teleost spermatozoa. Mitochondrion. 2017;**34**:49-55

[73] Somlo M, Cosson J, Clavillier L, Krupa M, Laporte I. Identity problems concerning subunits of the membrane factor of the mitochondrial ATPase of *Saccharomyces cerevisiae*. European Journal of Biochemistry. 1982;**122**:369-374

[74] Dreanno C, Cosson J, Suquet M, Dorange G, Fauvel C, Cibert C, et al. Effects of osmolality, morphology and intracellular nucleotide content during the movement of sea bass (*Dicentrarchus labrax*) spermatozoa. Journal of Reproduction and Fertility. 1999;**116**:113-125

[75] Terner C, Korsh G. The oxidative metabolism of pyruvate, acetate and glucose in isolated fish spermatozoa. Journal of Cellular and Comparative Physiology. 1963;**62**:243-249

[76] Christen R, Gatti JL, Billard R. Trout sperm motility: The transient movement of trout sperm is related to changes in the concentration of ATP following the activation of the flagellar movement. European Journal of Biochemistry. 1987;**166**:667-671

[77] Inoda T, Ohtake H, Morisawa M. Activation of respiration and initiation of motility in rainbow trout spermatozoa. Zoological Science. 1988;**5**:939-945

[78] Ingermann RL, Robinson ML, Cloud JG. Respiration of steelhead trout sperm: Sensitivity to pH and carbon dioxide. Journal of Fish Biology. 2003;**62**:13-23

[79] Dreanno C, Cosson J, Suquet M, Seguin F, Dorange G, Billard R. Nucleotide content, oxidative phosphorylation, morphology, and fertilizing capacity of turbot (*Psetta maxima*) spermatozoa during the motility period. Molecular Reproduction and Development. 1999;**53**:230-243

[80] Gosh RI. Energy metabolism of fish spermatozoids. A review. Gidrobiol Zh. 1989;**25**:61-71

[81] Billard R. La mobilité du spermatozoïde de poisson: Aspects énergétiques. In: As Jornadas Internationales de Reproduction animal. Zaragosa, Espagne; 1990. pp. 163-186

[82] Billard R, Cosson MP. The energetics of fish sperm motility. In: Gagnon C, editor. Controls of Sperm Motility: Biological and Clinical Aspects. Boca Raton, FL: CRC Press; 1990. pp. 153-173

[83] Dreanno C, Suquet M, Quemener L, Cosson J, Fierville F, Normand Y, et al. Cryopreservation of turbot (*Scophthalmus maximus*) sperm. Theriogenology. 1997;**48**:589-603

[84] Suquet C, Dreanno D, Dorange G, Normant Y, Quemener L, Gaignon JL, et al. The ageing phenomenon of turbot spermatozoa: Effects on morphology, motility and concentration, intracellular ATP content, fertilization, and storage capacities. Journal of Fish Biology. 2005;**52**:31-41. DOI: 10.1111/j.1095-8649.1998.tb01550.x

[85] Dreanno C, Seguin F, Cosson J, Suquet M, Billard R. H+-NMR and 31P-NMR analysis of energy metabolism of quiescent and motile turbot (*Psetta maxima*) spermatozoa. The Journal of Experimental Zoology. 2000;**286**:513-522

[86] Boryshpolets S, Dzyuba B, Stejskal V, Linhart O. Dynamics of ATP and movement in Eurasian perch (*Perca fluviatilis* L.) sperm in conditions of decreasing osmolality. Theriogenology. 2009;**72**:851-859

[87] Burness G, Schulte-Hostedde AI, Casselman SJ, Moyes CD, Montgomerie R. Sperm swimming speed and energetics vary with sperm competition risk in bluegill (*Lepomis macrochirus*). Behavioral Ecology and Sociobiology. 2004;**56**:65-70

[88] Burness G, Moyes CD, Montgomerie R. Motility, ATP levels and metabolic enzyme activity of sperm from bluegill (*Lepomis macrochirus*). Comparative Biochemistry and Physiology. 2005;**Part A 140**:11-17

[89] Robitaille PM, Munfort K, Brown G. 31P nuclear magnetic resonance study of trout spermatozoa at rest, after motility, and during short-term storage. Biochemistry and Cell Biology. 1987;**65**:474-485

[90] Saudrais C, Fierville F, Loir M, Le Rumeur E, Cibert C, Cosson J. The use of phosphocreatine plus ADP as energy source for motility of membrane-deprived trout spermatozoa. Cell Motility and the Cytoskeleton. 1998;**41**:91-106

[91] Perchec G, Chauvaud L, Suquet M, Cosson J, André F, Billard R. Changes in the movement characteristics and ATP content in the sperm of carp and turbot (teleost fishes). Comptes Rendus de l'Academie d'Agriculture de France. 1993;**6**:117-126

[92] Tsvetkova LI, Cosson J, Linhart O, Billard R. Motility and fertilizing capacity of fresh and frozen-thawed spermatozoa in sturgeons (*Acipenser baeri* and *A. ruthenus*). Journal of Applied Ichthyology. 1996;**12**:107-112

[93] Billard R, Cosson J, Fierville F, Brun R, Rouault T, Williot P. Motility analysis and energetics of the Siberian sturgeon *Acipenser baeri* spermatozoa. Journal of Applied Ichthyology. 1999;**15**:199-203

[94] Zietara MS, Slonimska E, Swierczynski J, Rurangwa E, Ollevier F, Skorkowski EF. ATP content and adenine nucleotide catabolism in African catfish spermatozoa stored in various energetic substrates. Fish Physiology and Biochemistry. 2004;**30**:119-127

[95] Cosson J. ATP: The sperm movement energizer. In: Kuester E, Traugott G, editors. Adenosine Triphosphate: Chemical Properties, Biosynthesis and Functions in Cells. New York: Nova Publisher Inc. 2012; pp. 1-46

[96] Ingermann RL. Energy metabolism and respiration in fish spermatozoa. In: Alavi SMH, Cosson JJ, Coward K, and Rafiee G, editors. Fish Spermatology. Oxford: Alpha Science International Ltd; 2008. pp. 241-266. ISBN: 978-1-84265-369-2

[97] Gibbons IR. Cilia and flagella of eukaryotes. The Journal of Cell Biology. 1981;**91**:107-124

[98] Tombes RM, Shapiro BM. Metabolite channeling: A phosphocreatine shuttle to mediate high energy phosphate transport between sperm mitochondrion and tail. Cell. 1985;**41**:325-334

[99] Tombes RM, Brokaw CJ, Shapiro BM. Creatine kinase dependent energy transport in sea urchin spermatozoa. Flagellar wave attenuation and theoretical analysis of high energy phosphate diffusion. Biophysics Journal. 1987;**52**:75-86

[100] Fedorov P, Grabic R, Fedorova G, Cosson J, Boryshpolets S, Dzyuba B. Development and application of LC/HRPS for quantification of adenine nucleotides, creatine phosphate, and creatine in sturgeon spermatozoa. Czech Journal of Animal Science. 2017;**62**:67-74

[101] Brokaw CJ. Nonsinusoidal bending waves of sperm flagella. The Journal of Experimental Biology. 1965;**43**:155

[102] Dadras H, Dzyuba B, Cosson J, Golpour A, Siddique MAM, Linhart O. Effect of water temperature on the physiology of fish spermatozoon function: A brief review. Aquaculture Research. 2017;**48**:729-740

[103] Ishimoto K, Cosson J, Gaffney EA. A simulation study of sperm motility hydrodynamics near fish eggs and spheres. Journal of Theoretical Biology. 2016;**389**:187-197. DOI: 10.1016/j.jtbi.2015.10.013

[104] Butts IAE, Prokopchuk G, Kašpar V, Cosson J, Pitcher TE. Ovarian fluid impacts flagella beating and biomechanical metrics of sperm between alternative reproductive tactics. The Journal of Experimental Biology. 2017;**220**:2210-2217. DOI: 10.1242/jeb.154195

[105] Siddique MAM, Cosson J, Psenicka M, Linhart O. A review of the structure of sturgeon egg membranes and of the associated terminology. Journal of Applied Ichthyology. 2014;**30**:1-10. ISSN: 0175-8659

[106] Boryshpolets S, Kowalski RK, Dietrich GJ, Dzyuba B, Ciereszko A. Different computer- assisted sperm analysis (CASA) systems highly influence sperm motility parameters. Theriogenology. 2013;**80**:758-765

[107] Rurangwa E, Kime DE, Ollevier F, Nash JP. The measurement of sperm motility and factors affecting sperm quality in cultured fish. Aquaculture. 2004;**234**:1-28

[108] Wilson-Leedy JG, Ingermann RL. Development of a novel CASA system based on open source software for characterization of zebrafish sperm motility parameters. Theriogenology. 2007;**67**(3):661-672

[109] Wilson-Leedy JG, Kanuga MK, Ingermann RL. Influence of osmolality and ions on the activation and characteristics of zebrafish sperm motility. Theriogenology. 2009;**71**:1054-1062

[110] Purchase CF, Earle PT. Modifications to the IMAGE J computer assisted sperm analysis plugin greatly improve efficiency and fundamentally alter the scope of attainable data. Journal of Applied Ichthyology. 2012;**28**:1013-1016

[111] Crimaldi JP, Browning HS. A proposed mechanism for turbulent enhancement of broadcast spawning efficiency. Journal of Marine Systems. 2004;**49**:3-18

[112] Mead KS, Denny MW. The effects of hydrodynamic shear stress on fertilization and early development of the purple sea urchin Strongylocentrotus purpuratus. The Biological Bulletin. 1995;**188**:46-56

[113] Katz DF, Blake JR, Pavieri-Fontana SL. On the movement of slender bodies near plane boundaries at low Reynolds number. Journal of Fluid Mechanics. 1975;**72**:529-540

[114] Cosson J, Huitorel P, Gagnon C. How spermatozoa come to be confined to surfaces. Cell Motility and the Cytoskeleton. 2003;**54**:56-63

[115] Gray J. The movement of sea-urchin spermatozoa. The Journal of Experimental Biology. 1955;**32**:775-801

[116] Farley GS. Helical nature of sperm swimming affects the fit of fertilization-kinetics models to empirical data. The Biological Bulletin. 2002;**203**:51-57

[117] Shaliutina A, Gazo I, Cosson J, Linhart O. Comparison of oxidant and antioxidant status of seminal plasma and spermatozoa of several fish species. Czech Journal of Animal Science. 2013;**58**:313-320

[118] Gazo I, Shaliutina-Kolešová A, Dietrich MA, Linhartova P, Cosson J. The effect of reactive oxygen species on motility parameters, DNA integrity, tyrosine phosphorylation and phosphatase activity of common carp (Cyprinus carpio L.) spermatozoa. Molecular Reproduction and Development. 2015;**82**(1):48-57. DOI: 10.1002/mrd.22442

[119] Shaliutina-Kolešová A, Gazo I, Cosson J, Linhart O. Protection of common carp (Cyprinus carpio L.) spermatozoa motility under oxidative stress by antioxidants and seminal plasma. Fish Physiology and Biochemistry. 2014;**40**(6):1771-1781. DOI: 10.1007/s10695-014-9966-z

[120] Dzyuba V, Cosson J, Dzyuba B, Rodina M. Oxidative stress and motility in tench, Tinca tinca, spermatozoa. Czech Journal of Animal Science. 2015;**60**:250-255

[121] Shaliutina A, Cosson J, Lebeda I, Gazo I, Shaliutina O, Dzyuba B, et al. The influence of cryoprotectants on sterlet Acipenser ruthenus sperm quality, DNA integrity, antioxidant responses, and resistance to oxidative stress. Animal Reproduction Science. 2015;**159**:66-76

[122] Avarre JC, Cosson J, Dugué R, Durand JD, Guinand B, Legendre M, et al. Gene expression in the testis of the male black-chinned tilapia Sarotherodon melanotheron heudelotii: Effect of acclimation to salinity. Peer

Journal. 2014;**2**:e702. DOI: 10.7717/peerj.702

[123] Shaliutina-Kolešová A, Sterba J, Kotas P, Rodina M, Cosson J, Linhart O. Protein profile of seminal plasma and functionality of spermatozoa during the reproductive season in carp Cyprinus carpio and rainbow trout Oncorhynchus mykiss. Molecular Reproduction and Development. 2016;**83**:968-982. DOI: 10.1002/mrd.22737

[124] Gazo I, Dietrich MA, Prulière G, Shaliutina-Kolešová A, Shaliutina O, Cosson J, et al. Protein phosphorylation in spermatozoa motility of Acipenser ruthenus and Cyprinus carpio. Reproduction. 2017;**154**(5):653-673. DOI: 10.1530/REP-16-0662

[125] Dumorné K, Figueroa E, Cosson J, Lee-Estevez M, Ulloa P, Valdebenito I, et al. Protein phosphorylation and ions effects on salmonid sperm motility activation. Reviews in Aquaculture. 2018;**10**:I727-I1737. DOI: 10.1111/raq.12197

[126] Boccaletto P, Siddique MAM, Cosson J. Proteomics: A valuable approach to elucidate spermatozoa post testicular-maturation in the endangered Acipenseridae family. Animal Reproduction Science. 2018;**192**:18-27. DOI: 10.1016/j.anireprosci.2018.03.033

[127] Chen DTN, Heymann M, Fraden S, Nicastro D, Zvonimir DZ. ATP consumption of eukaryotic flagella measured at a single-cell level. Biophysical Journal. 2015;**109**:2562. DOI: 10.1016/j.bpj.2015.11.003

[128] Fedorov P, Dzyuba B, Fedorova G, Grabic R, Cosson J, Rodina M. Quantification of adenosine triphosphate, adenosine diphosphate, and creatine phosphate in sterlet Acipenser ruthenus spermatozoa during maturation. Journal of Animal Science. 2015;**93**:1-8. DOI: 10.2527/jas.2015-9144

[129] Ellington WR, Kinsey ST. Functional and evolutionary implications of the distribution of phosphagens in primitive-type spermatozoa. The Biological Bulletin. 1998;**195**(3):264-272. DOI: 10.2307/1543138

Section 3

Climate Change and Aquatic Environment

Chapter 3

Climate Change and Monitoring of Temperate Lakes

Arne N. Linløkken

Abstract

Provided the predicted 2°C temperature increase during this century, lake ecology will go through dramatic changes, and this must be addressed in fish management in purpose of exploitation as well as in species preservation. In temperate lakes with fish communities dominated by cold-water and cool water fish, temperature increase will affect the species dominance. Extended growth season will benefit recruitment of less cool adapted species, total fish density may increase and growth will decrease of some species. Lakes dominated by salmonid fish may become dominated by cyprinids and percids. Primary production will increase due to extended growth season and increased precipitation. This can reduce the oxygen level in the deep layer of lakes when the organic matter decomposes, whereas the upper layer is too warm for cold-water species. In addition, increased density of small plankton feeding fish will reduce the algae feeding zooplankton. Lakes should be monitored by means of modern and sophisticated methods, monitoring lakes from satellites and in situ loggers, and pelagic fish may be counted by echosounding. To counteract increasing density of plankton feeding fish, fish biomass removal is a possible measure, though the effect is limited in time.

Keywords: pelagic zone, plankton, plankton feeders, predation, echo sounding, electronic loggers, remote sensing

1. Introduction

Research on biological resources of any kind may be divided in two categories: research on economical important species to attain optimal exploitation strategy and research to reveal ecological roles of species, commonly in a conservation perspective. The first point is obviously linked to the second, whereas the second point's connection to economy may be unclear and absent, at least apparently. Aquatic organisms of economic importance are mostly found in marine environments [1], while the economical value of freshwater fisheries in most cases is small [2], due to small water bodies and consequently small amounts of potential yield, though there are exceptions, like in the big lakes in North America [3]. It is also still important for hunting and gathering people in some parts of the world [4]. Freshwater bodies are nevertheless important to human society for many purposes, for drinking water, irrigation, and bathing and as landscape elements [5]. The water quality is therefore important in several perspectives, and guidelines for monitoring are recommended by the European Commission [6], among others.

IntechOpen

Water quality monitoring is concerned about water chemistry, physics, and biology. Certain elements are more important and are easy to monitor regularly, like phosphorus, nitrogen, acidity, clarity (Secchi depth, color, and turbidity), algae, and chlorophyll [6]. Optical instruments to analyze algae and chlorophyll concentrations are available, and a great benefit with the electronic measure devices is the possibility to install loggers for continuing monitoring [7]. Some biological elements are more laborious and expensive, like algae taxonomy, macro vegetation, zooplankton taxonomy, and fish abundance and ecology.

Lake ecology is affected by natural elements like local geology, climate, lake size, and bathygraphy [8], and a predicted climate change with a temperature increase of 2°C [9] during this century will provide serious consequences. In addition, human activity by the lake and in its catchment, even on remote places, affects the water quality. The organisms comprising a community may be more or less exclusive or rare, and some may be vulnerable or even threatened [10]. Ideally, there should be conducted genetic surveys on every species, but it would probably be a vast of resources. Nevertheless, species or populations that are considered as somehow vulnerable or unique should be analyzed by means of microsatellites or single-nucleotide polymorphism to describe the at-present state as a reference for future surveys [11–15]. Tissue samples should be preserved to be available for future analysis methods. The references are also useful for monitoring effective population size [13–15] and potential changes of allele frequencies, by random or as effects of natural selection.

2. Lakes as ecological indicators

Freshwater of lakes, rivers, and groundwater is important for all terrestrial life. There is scarcity of water in parts of the world [16], whereas in the temperate zone, it is usually available, though scarcity may occur in periods of the year. Water bodies are important sources of drinking water, and the quality is monitored for chemical and biological elements. Especially toxicants and pathogens are important, as they are mostly invisible by the eye. Other changes may be visual, like increased abundance of algae, planktonic, or on the bottom substrate or in fish nets. This may indicate eutrophication and could be serious. It indicates high levels of phosphorus, commonly the minimum factor in freshwater [8], though nitrogen also influences the primary production in lakes [17]. Increased phosphorus concentration may be added along lake shores or in the tributaries. Sources can be traced, and this could be a broken sewage pipe and runoff from a droppings pit or from fertilized fields resulting in an acutely polluted tributary. Algae bloom is a problem in hypertrophic lakes, and one characteristic trait is the cyanobacteria, among which, some may produce toxicants when occurring in high density [18, 19]. This has led to death of pasture cattle after drinking the water [20]. Even if such extremes are avoided, high algae production gain more biomass than the grazing chain from zooplankton to fish which may be consumed and decompose, and algae biomass is handled by the detritus chain in deepwater layer where oxygen deficit may occur [8]. Oxygen concentration at different depths is an important factor and is easy to measure with modern equipment. Oxygenated water precipitates iron (III) phosphorus and enriches the sediments, whereas oxygen deficit reduces the iron, and iron (II) phosphorus is soluble and is brought back into the water column during the spring mixing [8]. Phosphorus and chlorophyll a are therefore normally included in lake monitoring programs, and the chlorophyll a concentration is a good indicator of a lake's "health" condition [21]. Oxygen is very serious and may in worst case become chronic, and the lake sediments must be covered or replaced with other kinds of substrate, gravels and sand with low phosphorus content [22].

The lake characteristics and the fish community are mutually dependent of each other, but the fish community also depends on accessibility for freshwater species. Freshwater organisms immigrated after the glaciations, and the migrations were hampered or stopped by the topography. Mountains and waterfalls make up effective obstacles to fish distribution, and the distribution of species was determined by the time of immigration and the time of land uplift that created the obstacles. The distance from the glacial refuges and the topography decided the possibility for the entrance. Some fish communities therefore have few species, unaffected of the lake environment. This is clearly demonstrated in Norway, where the river systems in the western part of the country, that is, west of the mountain range, almost exclusively harbor fish of anadromous origin, which means no cyprinids, percids, or pike (*Esox lucius*) [23]. Cyprinid species normally dominate in eutrophic lakes, when present [24–26], and if not, a lake can be eutrophic but with a simpler community. It makes a difference whether the dominating pelagic fish species are cyprinids, coregonids, or Arctic charr (*Salvelinus alpinus*) [27]. They can all feed on zoo-plankton, but with different efficiency, due to different body size, population density and not least, different density of the gillrakers that filter food items from the water. The dense gill rakers of most of the cyprinids filter small zooplankton species that slip through the gill rakers of coregonids, not to mention those of Arctic charr and brown trout (*Salmo trutta*) [28]. This again affects the zooplankton grazing capacity, as the most important herbivorous species are large and more catchable than the smaller species, and algae blooms become more probable [29, 30].

Lakes may serve perfectly as indicators of environmental health condition of its surroundings and its catchment, and lakes are easily observable. Watercolor, clarity, and vegetation development can alarm the public in a lake's vicinity if dramatic changes occur. Bathing, fishing, or just observations from the shore may give a clue.

3. Forming of lakes and their ecosystems

The occurrence of a lake demands a substrate tight enough to hold water, some kind of damming, natural or manmade (eventually built by beavers), and a water supply that exceeds evaporation. Tectonic processes, land uplift, landslides, volcanoes, and quaternary processes can create holes capable of holding water when filled [8]. On the Northern Hemisphere, the glaciers, until 10,000–12,000 years ago, caused land excavations that became lakes and moraines that could dam the water.

In glaciated areas, when the ice thawed, and a completely barren land appeared, the primary succession started, affected by the local geology, and minerals bound in rocks and stones were released by physical and chemical weathering. Algae, lichens, mosses, and plants started assimilating CO_2 and, according to their needs, phosphorus, nitrogen, and other minerals [8]. Primary production was not hampered by organic bound nutrition and could flourish from the beginning. Grazers appeared and nutrition became in part bound in biota. Lakes got their sediments, slowly but steadily increasing, littoral macro vegetation developed, and animals, crustaceans, insects, mollusks, and others, appeared. Fish were more limited by waterways than most other aquatic organisms, like insects and small invertebrates that can be carried by other organisms or even by wind. In elevated areas, fish were not necessarily a result of the natural process but in many cases brought there by humans [31–33]. This was probably the first significant impact man had on lakes. Stools from sparse populations of Stone Age humans had probably a nonsignificant effect on lake productivity. Introduction of fish, which are mostly second (or higher)-order consumers, adds a predation pressure on organisms of several phyla, classes, and orders, among which, arthropods (especially crustacean and insects), annelids, mollusks, vertebrates, and, among those, other fishes, are important. If one fish species or five or more fish species are present, they structure the community through predation, depending on the species assemblage and the physical and chemical prerequisites of the lake. When the EU Water Directive demands lakes to be restored/maintained in "good status" [34], this is not unambiguous as it depends on the original state, that is, what species, nutrition load, acidity, and content of dissolved particles are and were present.

Monitoring lakes should always be done by comparing the present state with an assumed original state, although a new state is not necessary disastrous. There are many examples of successful introductions of fish species, from a human point of view. What the original state was can to some extent be illuminated by means of sediment analysis, as organisms with scales, like planktonic crustaceans, are preserved [35, 36]. The state should be stable, if not in a mal state, and spring and autumn circulation turning the water column to supply the deep water with oxygen should take place regularly. That will keep the lake sound, so the detritus is treated effectively with oxygen present, and fish dying off under the ice is avoided, as well as bad smell from methane and hydrogen sulfide. Fishing and bathing are of interest for public, and this public use of environmental resources makes people conscious about their state.

4. Monitoring of lake ecology

When monitoring fish communities, observations and measurements are conducted with a certain precision, and most importantly, the methods are described thoroughly enough to be repeated by others. Monitoring the lake ecology, one way or the other, may be regarded as testing a very wide/imprecise hypothesis: Is something changing? A more precise hypothesis can be formulated later, if the monitoring reveals that something in fact is changing. One species may become more abundant, whereas others become sparse, and an explanation can hopefully be found among the factors or variables that were monitored. At present, the temperature is a hot candidate. Others are nutrition load, acidity, new (alien) species, and eventually new parasites [37, 38]. The latter may become very harmful to indigenous species and populations [39].

Important variables may be logged electronically and even be transferred wireless to data archives, and for the large-scale overview, remote sensing by means of satellite images is probably the optimal method [42]. The horizontal overview given by satellite images is superior to the traditional spot sampling. Images taken from planes may surpass the satellite images when it comes to sharpness and details but hardly when it comes to frequency and regularity. Satellite instruments record reflectance of electromagnetic waves of different wavelengths, primarily within the visual specter, blue, green and red color, but also infrared, and longer wave lengths. Reflectance of long waved radiation can be used to calculate temperature and humidity at the earth surface, terrestrial as well as aquatic [43]. Satellite surveillance can easily be done weekly, provided if the sky is not cloudy and the images are available on the Internet, many of them for free. By monitoring surface color, the concentration of chlorophyll a, suspended matter and Secchi depth can be estimated. If worrying values are observed, water samples may be taken to laboratory.

What topics, methods and experimental design should be recommended? During the last two to three decades, efforts are spent to reveal and predict effects of climate change, i.e., increase temperature and changed runoff patterns. Temperature trends alarmed scientists in the 1990s, and lake ecology was predicted to change if

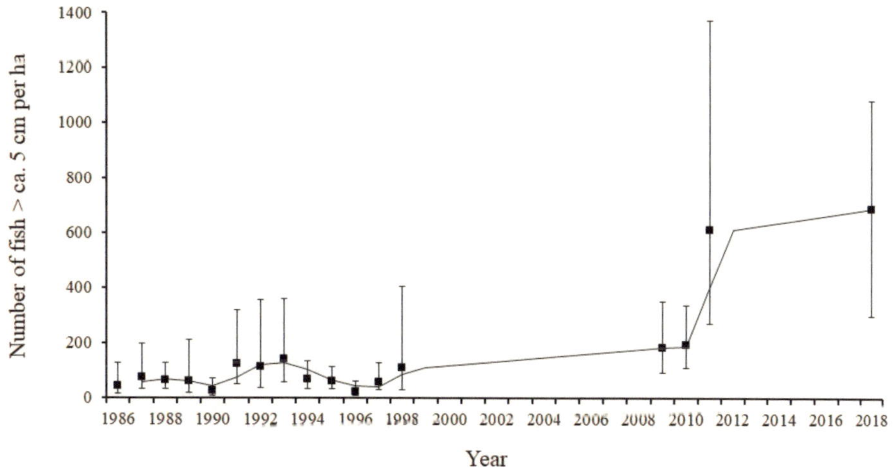

Figure 1.
Density of pelagic fish with body length ≥ approximately 5 cm recorded by means of a SIMRAD EY M echo sounder during 1986–2011 and by means of a SIMRAD EK 15 echo sounder in 2018 in Lake Osensjøen, Southeast Norway. Line shows moving average; vertical lines show 95% confidence intervals. For method description see Linløkken and Sandlund [40].

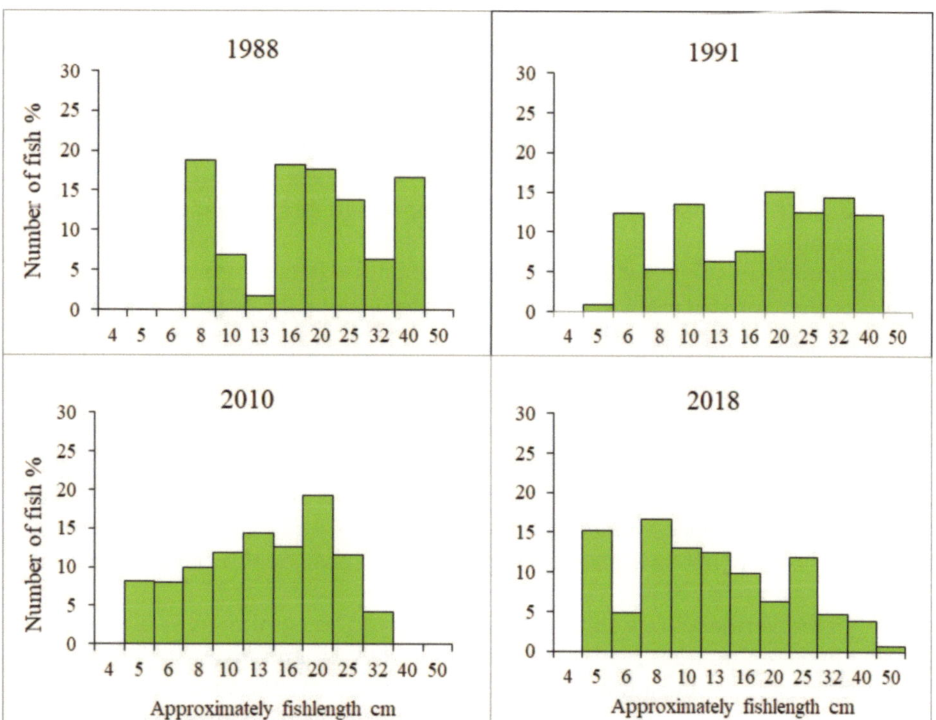

Figure 2.
*Length distribution of pelagic fish in four selected years (of **Figure 1**) based on echo target strength distribution from echo sounding in Lake Osensjøen, transformed to approximate fish lengths by the target strength—Fish length algorithm presented by Lindem and Sandlund [41].*

this tendency continued [44]. Time series of temperature, combined with existing knowledge about temperature effects on fish ecology, affecting recruitment success and growth [45–48], differing between species, will benefit some species and opposite to others. The algae community reflects the lake's nutrition load but is also affected by the grazing zooplankton, which is affected by the zooplankton feeding fish [27]. Coregonids are important plankton feeders in cool and cold-water lakes, whereas cyprinids compete more efficiently as temperature and nutrition load increase and oxygen concentration decreases, and this is the most frequent in shallow lakes [8]. How can this be monitored and what changes can be observed so far?

In the regulated and oligotrophic Lake Osensjøen in Southeast Norway, pelagic fish density, mainly whitefish and vendace, has been monitored by means of echo sounding since the 1980s, and during the first decade of this century, a pronounced density increase was observed, and a study was published in 2015 [40]. Year-class strength of both species was positively correlated with summer temperature, especially for vendace, as whitefish recruitment was also affected by the regulated water level (due to spawning sites at relatively shallow water). The vendace population increased, whereas whitefish in the pelagic zone decreased. A follow-up survey in 2018 confirmed high density of pelagic fish, and there was more than a sixfold increase since the 1980s and 1990s (**Figure 1**), and simultaneously, the proportion of fish >30 cm decreased, and the proportion of fish <20 cm increased (**Figure 2**). This has probably had an effect on the zooplankton community, of which samples are collected, but so far not analyzed. As the density increase was due to increased density of vendace, the growth rate of vendace has decreased [40].

5. Monitoring of lakes using satellites

Eutrophication and temperature increase may be expected to cause cyprinid dominance and amplify the effects of eutrophication through more intensive grazing on zooplankton. Temperature increase also affects recruitment and growth of percids positively [46, 48–50], among which perch and pikeperch are piscivorous and may play an important role in regulation of roach (*Rutilus rutilus*) and other cyprinids [51]. Exploitation with gill net fishing should be performed with great care to retain a sufficient number of large predatory specimens. What sufficient means should be a subject of research, which should otherwise concentrate on the description of species biomass of aquatic biotopes, with special attention on functional groups; who is eating who or what? Can it be stated an optimal balance between biomasses of consumers of first and second/third order, like between omnivorous/benhivorous/planktivorous cyprinids and species of higher trophic levels, like perch, pikeperch (*Sander lucioperca*) and pike and eventually the piscivorous cyprinid, the asp (*Aspius aspius*)? If not, serious measures may become necessary, like removal of fish biomass, which is shown to have a positive effect on zooplankton abundance, though not long lasting [52–54]. Stocking top predators, affecting the zooplanktivorous fish, may also have a positive effect on herbivorous zooplankton abundance [51, 55].

To get an overview, lakes and river systems should be monitored by means of satellite images. These are available from several sources, and many are for free [43]. The resolution of the images varies, and whereas some satellites, like the Sentinel 3 A and B satellites, have rather low resolution with 300 × 300 m pixels in the visual wave specter, the two twin satellites pass every second day, that is, together they collect daily images of every spot of the inhabited world [56]. The Landsat 7 and 8 satellites depict an area every eighth day [43], and the Sentinel 2 A and B satellites do it every fifth day [57]. These pairs of satellites have pixels of 30 × 30 m and

10 × 10 m, respectively, and the images are useful to characterize lakes with 5–10 km^2 surface area. The images consist of different color bands, and these bands, and combinations of them, may be used to develop algorithms for environmental factors like chlorophyll a [43], if combined with in situ measurements. The distribution of colors nevertheless may give a clue of horizontal variation of, for example, chlorophyll.

In the southernmost East Norway, lakes of the Halden river system exhibits pronounced variation of trophic state. A satellite image of August 13, 2018, from Landsat 8, with manipulated pseudo colors showing the relationship between the reflectance of the green (reflected by chlorophyll) and the red (absorbed by chlorophyll) color bands, shows colors from blue to green, yellow, and orange/red (**Figure 3**).
In the Lake Bjørklangen and Lake Hemnessjøen, the chlorophyll a concentration has through the years frequently been measured to 10–15 μg/l; in Lake Rødenessjøen it has been measured to 5–10 μg/l [58], whereas in the "blue" Lakes Setten and Rømsjøen [59], the chlorophyll a concentration is <5 μg/l. It must be added that the pixel values are also influenced by turbidity, that is, suspended solids, like in

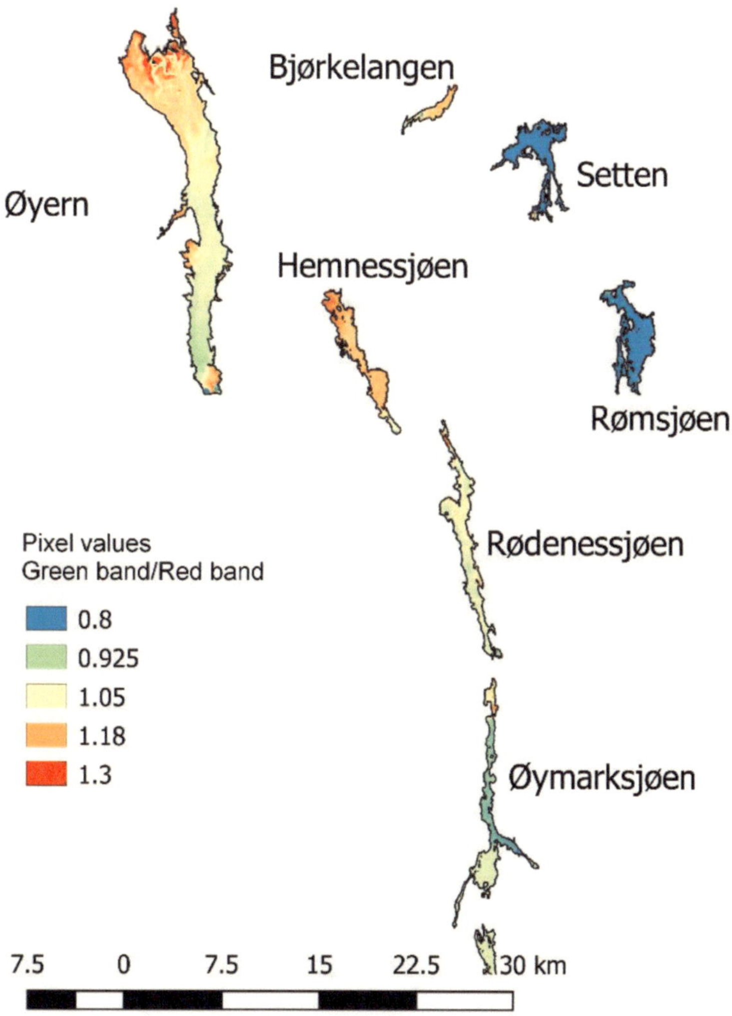

Figure 3.
Seven lakes in southernmost East Norway (Lat/Lon: 59°20′32″–59°56′60″, 11°02′13″–11°48′22″) showing horizontal variation of the ratio (pixel values of the green color band)/(pixel values of the red color band), based on a Landsat 8 satellite image, roughly indicating the distribution of chlorophyll [60].

the inlet of Lake Øyern and in the neighboring Glomma river system, which is not corrected for in this presentation.

In situ sampling, varying in space (horizontal) and time (seasonal), should be combined with satellite image analysis, to estimate reliable algorithms for the relationships between the chlorophyll a concentration, clarity and turbidity, and reflectance values (corrected pixel values) of different color bands of the satellite images. When this is established, satellite images can be used for regularly monitoring, daily in large lakes (>50 km^2) and weekly in smaller lakes, though it will depend on the weather, that is, the cloudiness, which must be minor and surely not cover the lake. This can reveal point sources of pollution, and in situ sampling may be conducted during few days.

Fish monitoring is commonly done by gill net fishing or trawling, laborious and costly, and therefore is not conducted too often. As the pelagic fish stock is of special interest in eutrophication, hydro acoustic equipment is recommendable, and that can be done regularly, like every or every third or fifth year. The time of year and time of day affect the results, due to the spatial distribution of the fish, which must be in the pelagic zone and not too close to the surface to be recorded. An echo sounder counts the single fish, integrates schools to numbers of fish, and estimates density and target strength distribution which can be transformed to fish size distribution. This method can easily record increased density and changed size distribution, which is a probable result of climate change and is possibly the case in the Lake Osensjøen (**Figures 1** and **2**).

6. Conclusion

Lakes play important roles as ecological elements in nature and comprise important resources for human societies. They are also important as indicators of the ecological state of their catchment and are relatively easy to observe and monitor, by in situ sampling, by electronic logging, and by means of remote sensing from satellites. Thorough studies should be conducted to describe the at-present state of lakes, which is now largely taken care of through the EU Water Frame Directive, and it should be linked to monitoring routines by electronic logging devices and to remote sensing by means of satellite images. Algorithms relating reflectance to situ measurements should be worked out on a broad scale.

Acknowledgements

This chapter was supported by Hedmark University College/Inland Norway University College, and Glommen and Laagen Brukseierforening.

Conflict of interest

There is no conflict of interest.

DOI: http://dx.doi.org/10.5772/intechopen.84393

Author details

Arne N. Linløkken
Inland Norway University of Applied Sciences, Hamar, Norway

*Address all correspondence to: arne.linlokken@inn.no

IntechOpen

© 2019 The Author(s). Licensee IntechOpen. This chapter is distributed under the terms of the Creative Commons Attribution License (http://creativecommons.org/licenses/by/3.0), which permits unrestricted use, distribution, and reproduction in any medium, provided the original work is properly cited.

References

[1] Anonymous. Oceans, Fisheries and Coastal Economies: World Bank Groups; 2018. Available from: http://www.worldbank.org/en/topic/environment/brief/oceans

[2] Anonymous. Freshwater Fishing. Economic Impact 2016: Climate Central; 2018. Available from: http://www.climatecentral.org/gallery/maps/economic-impacts-of-freshwater-fishing

[3] Commision GLF. Ann Arbor, MI. Available from: http://www.glfc.org/the-fishery.php

[4] Redwood DG, Ferucci ED, Schumacher MC, Johnson JS, Lanier AP, Helzer LJ, et al. Traditional foods and physical activity patterns and associations with cultural factors in a diverse Alaska native population. International Journal of Circumpolar Health. 2008;**67**(4):335-348

[5] Baver S. Top 10 Uses of Lakes. 2018. Available from: http://www.yourarticlelibrary.com/geography/lakes/top-10-uses-of-lakes/77538

[6] Anonymous. Guidance on Surface Water Chemical Monitoring under Water Framework Directive. Luxemburg: European Commission; 2009. Report No.: 25-2009

[7] Kellner K, Posnicek T, Brandl M. An integrated optical measurement system for water quality monitoring. Procedia Engineering. 2014;**87**:1306-1309

[8] Ruttner F. Fundamentals of Limnology. 3rd ed. Toronto: University of Toronto Press; 1975

[9] EPA. Climate Change Science: United States Environmental Protection; 2017. Available from: https://19january2017snapshot.epa.gov/climate-change-science/future-climate-change_.html#Temperature

[10] EUC. European Red List: European Commission, Environment; 2018. Available from: http://ec.europa.eu/environment/nature/conservation/species/redlist/index_en.htm

[11] Souza-Shibatta L, Kotelok-Diniz T, Ferreira DG, Shibatta OA, Sofia SH, de Assumpção L, et al. Genetic diversity of the endangered neotropical cichlid fish (*Gymnogeophagus setequedas*) in Brazil. Frontiers in Genetics. 2018;**9**:13

[12] Frankham R, Bradshaw CJA, Brook BW. Genetics in conservation management: Revised recommendations for the 50/500 rules, red list criteria and population viability analyses. Biological Conservation. 2014;**170**:56-63

[13] Linløkken AN, Haugen TO, Mathew PK, Johansen W, Lien S. Comparing estimates of number of breeders Nb based on microsatellites and single nucleotide polymorphism of three groups of brown trout (*Salmo trutta L.*). Fisheries Management and Ecology. 2016;**23**(2):152-160

[14] Wang J. A new method for estimating effective population sizes from a single sample of multilocus genotypes. Molecular Ecology. 2009;**18**(10):2148-2164

[15] Waples RS, Luikart G, Faulkner JR, Tallmon DA. Simple life-history traits explain key effective population size ratios across diverse taxa. Proceedings of the Royal Society B-Biological Sciences. 2013;**280**(1768)

[16] Anonymous. Water Supply: The World Bank Group; 2018. Available from: https://www.worldbank.org/en/topic/watersupply

[17] Canham CD, Pace ML, Weathers KC, McNeil EW, Bedford BL, Murphy L, et al. Nitrogen deposition and lake nitrogen concentrations: A regional

analysis of terrestrial controls and aquatic linkages. Ecosphere. 2012;**3**(7):art66

[18] Sinclair JL, Hall S, Berkman JA, Boyer G, Burkholder J, Burns J, et al. Occurrence of cyanobacterial harmful algal blooms workgroup report. Advances in Experimental Medicine and Biology. 2008;**619**:45-103

[19] Lopez CB, Jewett EB, Dortch Q, Walton BT, Hudnell HK. Scientific Assessment of Freshwater Harmful Algal Blooms. Interagency Working Group on Harmful Algal Blooms, Hypoxia, and Human Health. Council on Environmental Quality Office of Science and Technology Policy Executive Office of the President. Washington DC; 2008

[20] Puschner B, Galey FD, Johnson B, Dickie CW, Vondy M, Francis T, et al. Blue-green algae toxicosis in cattle. Journal of the American Veterinary Medical Association. 1998;**213**(11):1605, 571-1607

[21] Anonymous. National Aquatic Resource Surveys. Indicators: Chlorophyll a: United States Environmental Protection Agency; 2016. Available from: https://www.epa.gov/national-aquatic-resource-surveys/indicators-chlorophyll

[22] Bramm M, Christensen I. Management and Restoration of Lakes in Denmark. LakePromo: County of North Jutland Department of Aquatic Environment; 2006

[23] Huitfeldt-Kaas H. Ferskvandsfiskenes Utbredelse Og Indvandring i Norge: Med et tillæg Om Krebsen. Centraltrykkeriet: Kristiania; 1918

[24] Olin M, Rask M, Ruuhijarvi J, Kurkilahti M, Ala-Opas P, Ylonen O. Fish community structure in mesotrophic and eutrophic lakes of southern Finland: The relative abundances of percids and cyprinids along a trophic gradient. Journal of Fish Biology. 2002;**60**(3):593-612

[25] Persson L, Diehl S, Johansson L, Andersson G, Hamrin SF. Shifts in fish communities along the productivity gradient of temperate lakes—Patterns and the importance of size-structured interactions. Journal of Fish Biology. 1991;**38**(2):281-293

[26] Svärdson G. Interspecific population dominance in fish communities of Scandinavian lakes. Institute of Freshwater Research Drottningholm Report. 1977;**55**:144-172

[27] Hessen DO, Faafeng BA, Andersen T. Replacement of herbivorous zooplankton species along gradients of ecosystem productivity and fish predation pressure. Canadian Journal of Fisheries and Aquatic Sciences. 1995;**52**(4):733-742

[28] Jensen H, Kiljunen M, Knudsen R, Amundsen P-A. Resource partitioning in food, space and time between Arctic charr (*Salvelinus alpinus*), Brown trout (*Salmo trutta*) and European whitefish (*Coregonus lavaretus*) at the southern edge of their continuous coexistence. PLoS One. 2017;**12**(1):e0170582

[29] Sanni S, Wærvågen SB. Oligotrophication as a result of planktivorous fish removal with rotenone in the small, eutrophic lake Mosvatn, Norway. Hydrobiologia. 1990;**200**:263-274

[30] Thiel R. The impact of fish predation on the zooplankton community in a Southern Baltic Bay. 1996. 123-137

[31] Indrelid S. De første bosetterne. In: Barth EK, editor. Hardangervidda. Oslo: Luther Forlag; 1985. pp. 97-111

[32] Sønstebø JH, Borgstrøm R, Heun M. Genetic structure of brown

trout (*Salmo trutta* L.) from the Hardangervidda mountain plateau (Norway) analyzed by microsatellite DNA: A basis for conservation guidelines. Conservation Genetics. 2007;**8**(1):33-44

[33] Sønstebø JH, Borgstrøm R, Heun M. High genetic introgression in alpine brown trout (*Salmo trutta L.*) populations from Hardangervidda, Norway. Ecology of Freshwater Fish. 2008;**17**(1):174-183

[34] Anonymous. Brussels: The EU Water Frame Directive; 2018. Available from: https://publications.europa.eu/en/publication-detail/-/publication/ff6b28fe-b407-4164-8106-366d2bc02343/language-en/format-PDF/source-81652204

[35] Saulnier-Talbot É. Paleolimnology as a tool to achieve environmental sustainability in the anthropocene: An overview. Geosciences. 2016;**6**(2):26

[36] Gregory-Eaves I, Beisner BE. Palaeolimnological insights for biodiversity science: An emerging field. Freshwater Biology. 2011;**56**(12):2653-2661

[37] Idrisi N, Mills EL, Rudstam LG, Stewart DJ. Impact of zebra mussels (*Dreissena polymorpha*) on the pelagic lower trophic levels of Oneida Lake, New York. Canadian Journal of Fisheries and Aquatic Sciences. 2001;**58**(7):1430-1441

[38] Nicholls KH. Evidence for a trophic cascade effect on north-shore western Lake Erie phytoplankton prior to the zebra mussel invasion. Journal of Great Lakes Research. 1999;**25**:942-949

[39] Johnsen BO, Jenser AJ. The gyrodactylus story in Norway. Aquaculture. 1991;**98**(1):289-302

[40] Linløkken AN, Sandlund OT. Recruitment of sympatric vendace (*Coregonus albula*) and whitefish (*C. Lavaretus*) is affected by different environmental factors. Ecology of Freshwater Fish. 2015;**25**(4):652-663

[41] Lindem T, Sandlund OT. Ekkoloddregistreringer av pelagiske fiskebestander i innsjøer. Fauna. 1984;**37**:105-111

[42] Pause M, Schweitzer C, Rosenthal M, Keuck V, Bumberger J, Dietrich P, et al. In situ/remote sensing integration to assess forest health—A review. Remote Sensing. 2016;**8**(6):471

[43] U.S. Geological Survey. Landsat 8. Data Users Handbook. Sioux Falls: U. S. Geological Survey; 2016

[44] Lehtonen H. Potential effects of global warming on northern European freshwater fish and fisheries. Fisheries Management and Ecology. 1996;**3**:59-71

[45] LeCren ED. Observations on the growth of perch (*Perca fluviatilis*) over twenty-two years with special reference to effects of temperature and changes in population density. Journal of Animal Ecology. 1958;**27**:287-334

[46] Linløkken A, editor Temperature dependence of Eurasian perch (*Perca fluviatilis*) recruitment. Percis III: The Third International Percid Fish Symposium; Madison, Wisconsin: University of Wisconsin Sea Grant Institute; 2003

[47] Linløkken AN, Hesthagen T. The interactions of abiotic and biotic factors influencing perch *Perca fluviatilis* and roach *Rutilus rutilus* populations in small acidified boreal lakes. Journal of Fish Biology. 2011;**79**(2):431-448

[48] Tolonen A, Lappalainen J, Pulliainen E. Seasonal growth and year class strength variations of perch near the northern limits of its distribution range. Journal of Fish Biology. 2003;**63**(1):176-186

[49] Lehnert SJ, Pitcher TE, Devlin RH, Heath DD. Red and white Chinook salmon: Genetic divergence and mate choice. Molecular Ecology. 2016;**25**(6):1259-1274

[50] Neuman E. The growth and year class strength of perch in some Baltic archipelagos, with special reference to temperature. Report from Institute of Freshwater Research Drottningholm. 1976;**55**:51-70

[51] Findlay DL, Vanni MJ, Paterson M, Mills KH, Kasian SEM, Findlay WJ, et al. Dynamics of a boreal Lake ecosystem during a long-term manipulation of top predators. Ecosystems. 2005;**8**(6):603-618

[52] Faafeng BA, Hessen DO, Brabrand A, Nilssen JP. Biomass manipulation and food-web dynamics–The importance of seasonal stability. Hydrobiologia. 1990;**200**:119-128

[53] Jeppesen E, Meerhoff M, Jacobsen BA, Hansen RS, Søndergaard M, Jensen JP. Restoration of shallow lakes by nutrient control and biomanipulation—The successful strategy varies with lake size and climate. Hydrobiologia. 2007;**581**:269-285

[54] Jeppesen E, Søndergaard M, Lauridsen TL, Davidson TA, Liu Z, Mazzeo N. Biomanipulation as a restoration tool to combat eutrophication: Recent advances and future challenges. Advances in Ecological Research. 2012;**47**:411-488

[55] Brabrand A, Faafeng B. Habitat shift in roach (*Rutilus rutilus*) induced by pikeperch (*Stizostedion lucioperca*) introduction–predation risk versus pelagic behavior. Oecologia. 1993;**95**(1):38-46

[56] ESA. Sentinel-3 User Handbook: In: Agency ES, editor. European Commission, 150. 2013

[57] ESA. Sentinel-2 User Handbook: European Space Agency; 2015. 64 p

[58] Haande S, Rohrlack T, Kyle M. Utvikling av vannkvalitet i Haldenvassdraget. Sammenstilling av lange tidsserier (1968-2013). Oslo: Norwegian Institute of Water Research; 2014

[59] Bjørndalen K, Vallner THP. Rømsjøen—en vannfaglig vurdering. vol Rapport. Moss: Østfold County Environmental Administration; 1985

[60] Linløkken AN. Satellittovervåking av innsjøer—en metode for framtida? Vann. 2018;**53**(4):355-365

Section 4

Migration in Aquatic Species

Chapter 4

Migration, Dispersal, and Gene Flow of Harvested Aquatic Species in the Canadian Arctic

Ross F. Tallman, Steven H. Ferguson, Les N. Harris, Kevin J. Hedges, Kimberly L. Howland, Nigel E. Hussey, Marianne Marcoux, Cory J.D. Matthews, Zoya A. Martin and Jean-Sebastien Moore

Abstract

Migration occurs when key aspects of the life cycle such as growth, reproduction, or maintenance cannot all be completed in one location. The Arctic habitats are variable and Arctic species are often migratory. The predictable nature of migrations in both space and time allow Arctic people to harvest fishes and marine mammals. We describe migratory/dispersal behavior in four types of taxa from the Canadian Arctic: anadromous and freshwater fishes, marine fishes, marine invertebrates, and marine mammals. Patterns of migration are remarkably different between these groups, in particular between distances migrated, seasonal timing of migrations, and the degree of reproductive isolation. Migratory anadromous and freshwater fishes become adapted to specific locations resulting in complex life histories and intra- and inter-population variation. Marine mammals not only migrate longer distances but also appear to have distinct demographic populations over large scales. Marine fishes tend to be panmictic, probably due to the absence of barriers that would restrict gene flow. Migratory patterns also reflect feeding or rearing areas and/or winter refugia. Migratory patterns of harvested aquatic organisms in the Canadian north are extremely variable and have shaped the north in terms of harvest, communities, and culture.

Keywords: migration, genetics, harvest, ecology, Arctic, fish, marine mammal, invertebrate, distribution

1. Introduction

To grow, reproduce, and survive, organisms must find environments where they can successfully complete all aspects of their life cycle. In most taxa, the life cycle needs (e.g., foraging and reproduction) cannot be met in a single habitat. In such cases, the fitness of individuals benefits from movement to an alternate habitat [1]. If reproductive, rearing, foraging, or refugial environments are not all sympatric, then migrations are required [2]. As a result, many animals have evolved life history strategies that include coordinated movement from one habitat to another during specific life stages. This synchronous, directed movement of part or all of a population between discrete habitats is called "migration." In the Canadian Arctic, habitats can be relatively barren but some regions have great productivity (**Figure 1**). The "patchiness" of the Canadian Arctic encourages the evolution of migrations and the majority of endemic species are migratory to some degree.

Where resources are patchy in space and time, such as the Arctic, species must evolve migratory life histories for survival [3]. The evolution of migration can be interpreted in terms of a balance between benefits gained from migration and costs of migration [4]. Migration confers an advantage for finding the most suitable spawning habitat, more productive feeding areas, and/or finding refuge from inclement conditions [5, 6]. On the other hand, migration is costly in terms of mortality to both juveniles and adults and the energy used to migrate [7]. For the Arctic charr migrating downstream to the ocean, the cost occurs before the benefit of a richer food source. Phenotypes that can reduce the cost of migration will be favored [7]. For example, species have evolved adaptations in their morphology to facilitate more efficient swimming as well as osmoregulatory adaptations for moving between fresh and saltwater.

Among both plants and animals, dispersal usually takes place at the time of reproduction. Dispersal is defined as the movement of individual organisms from their birthplace to other locations for breeding [8]. Arctic aquatic organisms must

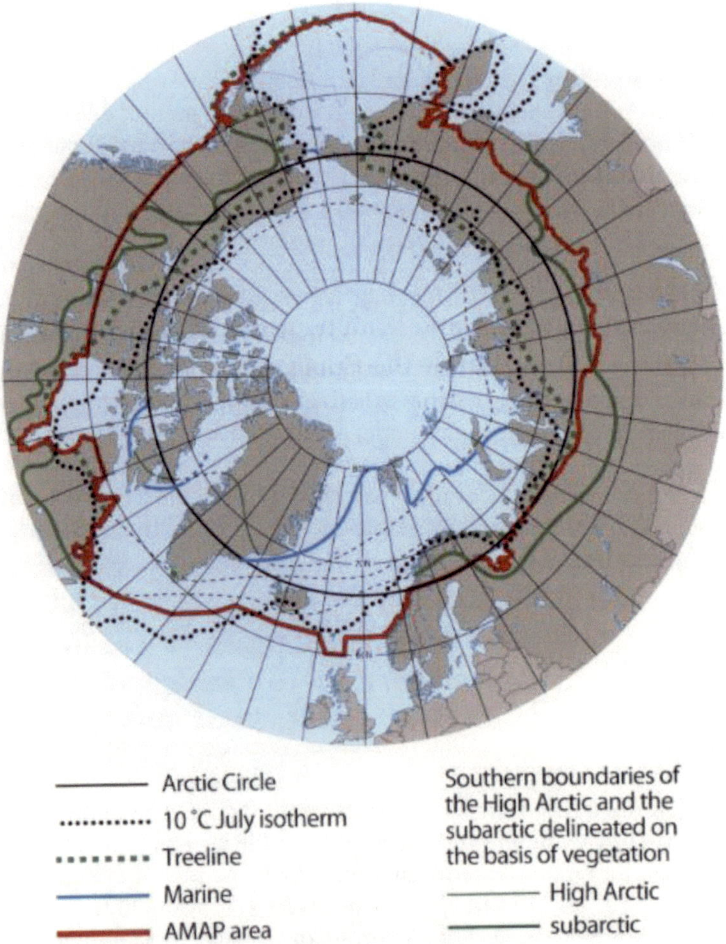

Figure 1.
Definitions of the Arctic region (source AMAP [122]). Comparisons between different definitions of the Arctic region are shown. The AMAP area is considered the most encompassing biologically for aquatic systems.

also be exceptional colonizers, and dispersal plays a critical role in the expansion of species in a post-glacial environment.

Until the last 100 years, the survival of humans in the Arctic was entirely dependent on the fauna of the region as evidenced by the strong hunting cultures that persist to this day. Understanding the movement patterns of aquatic species allowed exploitation of migratory species by the aboriginal societies. Migrations consistently brought the desired species in concentrations to particular locations. Currently, a variety of aquatic species from invertebrates to marine mammals are utilized by the Inuit and other Arctic people for subsistence and commercial purposes.

The Arctic represents a relatively untouched region for fishery exploitation, given its remote and logistically challenging location and inaccessibility due to land-fast ice [9]. The region is therefore considered promising for fishery development, especially with receding ice cover as a result of global climate change [9]. Given this potential, there are mounting concerns over fishery development related to the glaring lack of basic biological data for the majority of species and populations and our understanding of the ecosystem as a whole [9, 10].

2. What are the harvested species in the Canadian Arctic?

The Canadian Arctic is challenging to define and there are many possibilities. For this chapter, we use the definition provided by Arctic monitoring and assessment program (AMAP) (see **Figure 1**). The AMAP definition is a composite that reflects biological, geographical, and political definitions of the Arctic. The place names mentioned in the text are in **Figure 2**.

The list of harvested aquatic species in the Canadian Arctic is long and diverse and would require an exhaustive review to cover them all and we will only discuss a few key ones in this chapter. Freshwater fishes such as lake trout (*Salvelinus namaycush*), lake whitefish (*Coregonus clupeaformis*), and landlocked forms of Arctic charr (*S. alpinus*) are extremely important and are harvested throughout Arctic Canada by indigenous people of the region. The freshwater fishery in sub-Arctic Great Slave Lake, comprised of lake whitefish, lake trout, and inconnu (*Stenodus leucichthys*), is the largest commercial freshwater fishery north of 60 N latitude. At a community level, anadromous fishes comprise a large portion of the subsistence harvest across the Canadian Arctic. In Nunavut, anadromous Arctic charr are harvested commercially by most communities. Anadromous (those that migrate between fresh water where they spawn and/or overwinter to marine habitats where they feed) coregonines such as broad whitefish (*C. nasus*), inconnu, lake whitefish, least, and Arctic cisco (*C. sardinella* and *C. autumnalis*) are harvested during their fall spawning migrations along western Arctic coastal rivers and especially in the Mackenzie River where a fishery estimated at 150,000 kg/year is taken for subsistence purposes [11]. Northern anadromous Dolly Varden Char (*Salvelinus malma*) are mainly subsistence fished in the Beaufort Sea coastal waters [12] and along the Mackenzie River system on their return migration from the coast (e.g., Aklavik and Fort MacPherson). Anadromous Arctic charr are harvested throughout the territory of Nunavut in all coastal communities as well as in the communities of Paulatuk, Ulukhaktok, and Sachs Harbour in the Northwest Territories.

Marine fishes are harvested when available by many communities in the North. There are large industrial fisheries for northern shrimp (*Pandalus montagui* and *Pandalus borealis*) in Davis Strait and Hudson Strait, and for Greenland halibut (*Reinhardtius hippoglossoides*) in Baffin Bay and Davis Strait and in Cumberland Sound, Nunavut. The shrimp fishery harvests up to 35,000 metric tonnes annually.

Figure 2.
Map of Canada with northern place names relevant to the text. Communities are marked by a black dot. The Mackenzie River system is in blue.

Greenland halibut are widespread, live deep (200–2000 m) and are long-lived [13–19]. The Baffin Bay fishery is now the largest groundfish fishery on the east coast of Canada. The total harvest regularly exceeds 10,000 metric tonnes. The Cumberland Sound fishery is a community-based winter fishery (500 metric tonnes quota) executed by setting longlines from the flow edge of the winter ice pack or through holes cut in the landfast ice. Commercial fisheries on this species operate at Baffin Bay and Davis Strait [20, 21].

Marine mammals are an important harvested resource for coastal communities in the Arctic not only for subsistence purposes but also for commercial sale of Narwhal (*Monodon monoceros*) tusks and other derived products. Ringed seal (*Pusa hispida*), narwhal, beluga whale (*Delphinapterus leucas*), bowhead whale (*Balaena mysticetus*), and walrus (*Odobenus rosmarus*) are all important species that are harvested throughout Arctic Canada.

3. Why is understanding migration so important for harvesting and management?

Migratory pathways of Arctic animals bring them into vulnerable situations pathways bring them into vulnerable situations in time and space where humans can access them in a predictable manner. Because of this, the management is generally tied to the migration patterns and is frequently done on a community by community basis. As well, most harvesters are aboriginal and harvests are for both economic gain and food security. The exception is the offshore marine fisheries which are comparable to other large industrial fisheries around the world.

As well, migrating exposes the animals to an accumulation of stressors such as shipping, pollution, and blocked migration routes (e.g., streams drying up due to global warming) which all can dramatically reduce their fitness. Additionally, climate change is anticipated to affect the Canadian Arctic most severely causing a major change, principally an increase in extent and duration of the open-water season. The region has been heavily characterized by the presence of long seasonal ice cover and multiyear sea ice. In the future, the ice-free season will be much longer and areas of multiyear ice will diminish. The change will not only affect migration but may also allow the movement of non-Arctic species into the region, which may be competitors, predators, or forage items. Therefore, to manage fisheries, a thorough understanding of migration patterns of the harvested species is required. Migration patterns of harvested Canadian Arctic species have been under study for the last 20 years and considerable progress has been made in the last 5 years. Although remarkable strides have been made in the understanding of migration, there is still much to understand.

4. Migratory patterns

4.1 Freshwater and anadromous species

4.1.1 Coregonines

4.1.1.1 Inconnu (Stenodus leucichthys)

Inconnu have a partially circumpolar distribution. They mainly inhabit rivers flowing into the Arctic Ocean. They range from the Anderson River, NWT at latitude 128 W westward to 35 E on the edge of Finland. They are harvested throughout their range for subsistence and commercial purposes. Howland et al. [22–24] undertook a radio-telemetry, otolith microchemistry, and long-term seasonal gillnetting study which, along with a synthesis of existing historical data, revealed that both freshwater (those defined as only using riverine habitats and never migrating to the ocean) and anadromous Inconnu (*Stenodus leucichthys*) in the Mackenzie River, NT system are migratory. However, their feeding/overwintering habitats and the timing and distance of their spawning migrations differ substantially between life history types (**Figure 3**). Two factors, seasonal temperatures and distance of the migration, probably played a role in the timing of migration as the freshwater population migrated and spawned later by 2–2.5 months and 3 weeks, respectively, than did the anadromous population. There is little mixing between the anadromous forms in the lower Mackenzie River and freshwater Inconnu in the Great Slave Lake area. Tagging studies, however, have provided evidence for long distance movement in this species (~1800 km) suggesting the potential movement and mixing between the two forms [25]. Some fish seem to use only the upper river for their life cycle suggesting that a third riverine form occurs. In the lower Mackenzie River, Inconnu travel more broadly using coastal and the Mackenzie Delta for nourishment and as a winter refugium, whereas Inconnu in the Great Slave Lake area eat and survive the winter in the lake and migrate less distance to spawn. These results were reflected in differences in mitochondrial (Howland DFO unpublished) and microsatellite (Weins DFO unpublished) DNA analysis.

4.1.1.2 Broad whitefish (Coregonus nasus)

Broad whitefish have a partially circumpolar distribution. They mainly inhabit rivers flowing into the Arctic Ocean. They range from latitude 105 W westward to 50 E. They are harvested throughout their range for subsistence and commercial

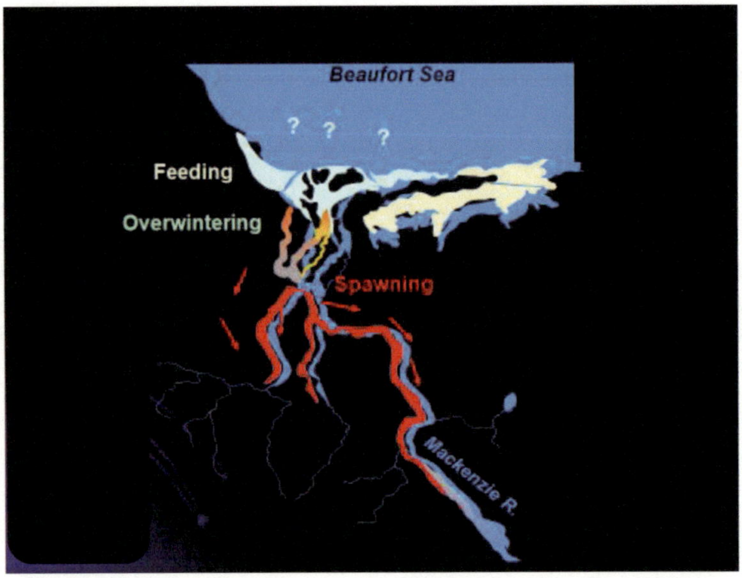

Figure 3.
The lower Mackenzie River showing spawning, overwintering, and feeding areas used by migratory Coregonids. The pale yellow or cream colored area is the suspected feeding area. The off-white area in the Mackenzie Delta is a proposed over-wintering area. The red shows the migration to spawning grounds. Blue is the water.

purposes in the lower Mackenzie River system of Canada's NWT. Broad whitefish exhibit three presumed life history strategies including anadromous, riverine, and lacustrine (complete their life cycles entirely within a lake) forms [26, 27]. One anadromous population of Broad whitefish is known to spawn in the Arctic Red River at Weldon Creek, about 160 miles upriver from the mouth [28]. The minimum total distance that Arctic Red River anadromous Broad whitefish migrate from their over-wintering grounds to spawning beds is 350–450 km. First-time spawners, arriving from rearing grounds on the Tuktoyaktuk Peninsula, migrate considerably farther [29] (**Figure 3**). Spawning locations for anadromous fishes have also been suggested in the Peel River and the mainstem Mackenzie River near the town of Fort Good Hope [29, 30]. Recently, Harris et al. [25] examined migratory strategies in this species using otolith microchemistry and documented at riverine form that undertakes migrations solely within the Peel River in the region.

A lacustrine form of Broad whitefish occurs in Travaillant Lake, a deep lake about 31km^2 in surface area [31]. Broad whitefish use the lake for rearing and overwintering. Reproduction occurs in two locations, in the outlet of Travaillant Lake above Andre Lake to the south and in a major inlet to Travaillant Lake, directly to the north [31, 32]. The migration distance to spawning grounds is short, only 5–12 km.

4.1.1.3 Arctic cisco (C. autumnalis) and least cisco (C. sardinella)

Least and Arctic cisco have a partially circumpolar distribution. They mainly inhabit rivers flowing into the Arctic Ocean. Least cisco range from latitude 100 W westward to 50 E. They are harvested throughout their range for subsistence purposes, but to our knowledge, they have not been fished commercially. Arctic cisco range from latitude 105 W westward to 40 E. They are harvested throughout their range for subsistence purposes and historically supported a large commercial fishery in Alaska [33, 34]. These species follow similar life styles with eggs hatching in spring and young-of-the-year are carried downstream with increased water flow from spring melt to coastal, brackish environments [33]. Least cisco distribute themselves in coastal brackish and fresh waters to feed in the summer months, but unlike Arctic cisco they do not migrate 400 km that separates Colville River, Alaska, and the Mackenzie Delta [33]. Young-of-the-year Least cisco and eastward Arctic cisco over-winter within areas of the Mackenzie Delta where there is a stable layer of freshwater under the sea ice [35]. Migration to natal streams occurs with the onset of sexual maturity. The age of sexual maturity in Siberian populations varies between 5 and 10 years but there are no studies in Canadian waters [35]. In contrast to Least cisco, Arctic cisco are distributed more widely within the coastal marine environment as a result of their ability to tolerate and acclimate to higher salinity [36]. A strong eastern wind facilitates the migration of Arctic cisco young-of-the-year to Alaska from the Mackenzie River system by wind-driven coastal currents that force the movement of warm, less saline water to Alaska [34]. Young-of-the-year that reach Alaskan waters reside in the area to forage and utilize the Colville River to over-winter. Upon reaching sexual maturity, Arctic cisco return to natal tributaries within the Mackenzie River system to spawn [37] (**Figure 3**).

Arctic cisco are thought to only spawn within four tributaries (Peel River, Arctic Red River, Liard River, and Great Bear River) in North America, which are part of the greater Mackenzie River system [38, 39]. There is strong evidence indicating that Arctic cisco have relatively long oceanic migrations of approximately 600–700 km, whereas Least cisco have shorter oceanic migrations of approximately 200–300 km [35, 40]. Presently, both species co-occur in the Arctic Red River, but little scientific research has been undertaken to understand their life history strategies.

4.1.1.4 Lake whitefish (C. clupeaformis)

Lake whitefish are located throughout most of Canada and Alaska. They are harvested throughout their range for subsistence and commercial purposes and form the largest freshwater fishery north of the 60 N in Great Slave Lake. Lake whitefish are considered relatively sedentary with migrations occurring within lakes over relatively short distances between feeding and spawning areas. Surprisingly, there is little information on the migration of Lake whitefish in Arctic or sub-Arctic regions. However, this little information shows a marked departure from the patterns in the southern part of their range. Lake whitefish in the Mackenzie River are considered unusual because they migrate along the river to coastal areas. Lake whitefish from the Mackenzie River estuary are believed to overwinter there and in the river's delta [41], and to spawn in various tributaries of the Mackenzie, such as the Rat, Peel, and Arctic Red rivers [42]. Lange and Tallman [43] noted that the number of ripe lake whitefish increased within the Arctic Red River in the months of September and October. Limited radio-tracking information suggest that they migrate similar distances to broad whitefish [35] (**Figure 3**).

4.1.2 Salmonines (Salvelinus spp.)

4.1.2.1 Dolly varden (Salvelinus malma)

The northern form of Dolly varden (*S.m. lordi*) is found in northwestern North America and northeastern Eurasia. In North America, populations range from Bristol Bay along the north slope of Alaska and the Yukon Territory, and east to the Mackenzie River [44]. Approximately 5–10% of the global population exists within Canadian waters. Population sizes are largely unknown.

Anadromous Dolly varden char (*Salvelinus malma*) are mainly fished for subsistence using gillnets in Beaufort Sea coastal waters of both Canada (Northwest Territories and Yukon) and Alaska [12]. They are also harvested in Inuvialuit and Gwich'in communities along the Mackenzie River system (e.g., Aklavik and Fort McPherson) on the return migration to headwater spawning/overwintering areas. Dolly varden have variable migration strategies [45–49]. For example, anadromous individuals may migrate to sea or part of the same population will stay in fresh water throughout its life; this behavior appears to be facultative and may be linked to early growth history [49] with residents being genetically indistinguishable from their anadromous counterparts [45]. There are also genetically distinct isolated freshwater populations above waterfalls in several river systems [50, 51]. Residents do not migrate to the sea, but remain in fresh water year-round. The majority of these are small males that adopt a 'sneak spawn' strategy (taking on a freshwater "resident" lifestyle) in order to fertilize eggs of depositing anadromous females [39]. While rare, a few cases of female residents have been documented in Canadian Arctic systems [46]. Residents are characterized by their small size, dark color, visible parr marks, and early maturation. Spawning of Dolly varden in Canadian rivers occurs from early September to late October just before freeze-up, with fry emerging from the gravel in May and June [39]. Anadromous juveniles remain in fresh water from 1 to 5 years before beginning annual migrations to feed in productive marine waters of the Beaufort Sea [49, 52, 53]. Downstream migration to the sea begins during spring freshet, which is typically in early to mid-June and may be linked with ice conditions in the Beaufort Sea [39, 47]. Upstream migration can begin as early as July and last until mid to October in some systems [43, 49]. Dolly varden of all life history stages show high fidelity to spawning/overwintering areas as evidenced by their distinct genetic structuring by river systems as well as tag return information [45, 51, 52, 54, 55]. Mature spawning adults tend to return to natal streams first, while smolts are usually the last fish to return [56]. Dolly varden are iteroparous, with populations/individuals typically migrating and spawning either annually or biennially [57, 58]; however, some populations are known to skip both migration and spawning in some years [54].

Gallagher et al. [59] used otolith strontium and multi-year mark-recapture information to characterize associations between migration patterns and spawning frequencies in an anadromous Dolly varden. They observed that fish either migrated annually after smoltification or periodically skipped an annual ocean migration to remain in fresh water and spawn. Annually migrating fish had lower longevity (≤9 years vs. ≤13 years). They also observed that some fish returned from the sea considerably earlier than the majority of other current-year migrants.

Based on recent studies involving the use of satellite tagging, migration in the marine environment can involve movements well offshore at least in the region of the Mackenzie estuary (Gallagher and Howland, unpublished) as well as off the coast of Alaska [48]. In general, Dolly varden have been noted to migrate longer distances than Arctic charr, for example, DeCicco [55, 60] recorded a migration from Alaska to Russia.

4.1.2.2 Arctic charr (S. alpinus)

Arctic charr (*Salvelinus alpinus*) have an unbroken circumpolar distribution. They are found in north flowing streams, rivers, and lakes around the Arctic Ocean. Arctic charr are important as an iconic symbol of the North, important for food security of aboriginal people, and form the basis for valued commercial fisheries. Most stocks are prosecuted using gillnets or weirs. Arctic charr have been recently studied for both dispersal patterns and for migrations. Dispersal can influence the process of local adaptation, when the dispersers successfully breed (i.e., gene flow occurs) in the non-natal habitat [61]. Fiords might be an important influence on gene flow and also may be a major barrier across Cumberland Sound, NU [62]. Arctic charr showed isolation by distance and it is thought that dispersal in Arctic charr follows a stepping stone pattern [62]. Anadromous Arctic charr display a complex migratory behavior throughout a large portion of their range. Breeding and nonbreeding individuals have very different movements, and for genetic analysis, it is important to distinguish them. As well, individuals do not necessarily reproduce every year, but to survive the winter, they must return to fresh water. Moore et al. [61] used a genetic assignment approach to study dispersal of charr from Cumberland Sound on Baffin Island, Canada. Dispersal estimates ranged from 15.8 to 25.5%, which is higher than recorded for other salmonids. Those returning to fresh water solely for overwintering and not spawning purposes were more likely to use non-natal habitats than breeding individuals, thus resulting in estimates of dispersal that overestimate the potential for gene flow among populations. Moore et al. [61] also parameterized a population genetic model showing that gene flow is probably sufficiently low to allow for local adaptation among populations. It is hypothesized that that Arctic charr home to their natal river to spawn, but may overwinter in rivers with the shortest migratory route to minimize the costs of migration in nonbreeding years [63]. Several studies in the Canadian Arctic have also suggested that dispersal is often asymmetrical among discrete stocks [61, 64] and that gene flow often conforms to a stepping-stone-pattern [55, 57]. These results underscore the importance of understanding patterns of dispersal for appropriately evaluating potential consequences for local adaptation and management.

In general, Arctic charr are not suspected to migrate long distances while summer foraging in the marine environment and that they tend to remain close to the shore [65–68]. More recently, Moore et al. [69] used an array of fixed acoustic receivers ($N = 42$) to track the summer marine movements of 121 anadromous Arctic charr equipped with acoustic transmitters at three locations in the Cambridge Bay region, where the largest commercial fishery for this species exists. They found that the seasonal time of movement between salt and fresh water depended on the river of origin rather than size or sex. The sexes differed in the distance they moved with the males moving further from where they were tagged. The fish remained in brackish estuarine environments for the most part and curiously travelled mainly westward from their river of origin. They appeared to be moving to the water in Wellington Bay that was the warmest and freshest. The pattern of movement was to rest in the estuaries of rivers and then move rapidly through the more marine areas. Charr preferred nearshore habitats based on the increased numbers of detections on receivers located less than 1.5 km from the coast. Finally, they noted an implication for fishery management because they observed evidence of extensive stock mixing throughout the summer, including at known fishing locations and periods (**Figure 4**). Mixed-stock migrations in marine habits appear to be a common phenomenon in this species [64, 70].

4.1.2.3 Lake trout (S. namaycush)

Lake trout are widespread in North America from the Arctic coast and near islands to the northern United States. They are harvested for subsistence, and sport throughout the north and commercially in Great Slave Lake.

Lake trout migrations are mainly limited to within lakes although since they inhabit some lakes such as Great Bear Lake that are freshwater oceans, these migrations may cover quite a distance (e.g., upwards of 60 km). Recently, Swanson et al. [71] noted that Lake trout could undertake limited migrations to the sea to take advantage of the lipid reaching prey sources present in marine habitats. Using

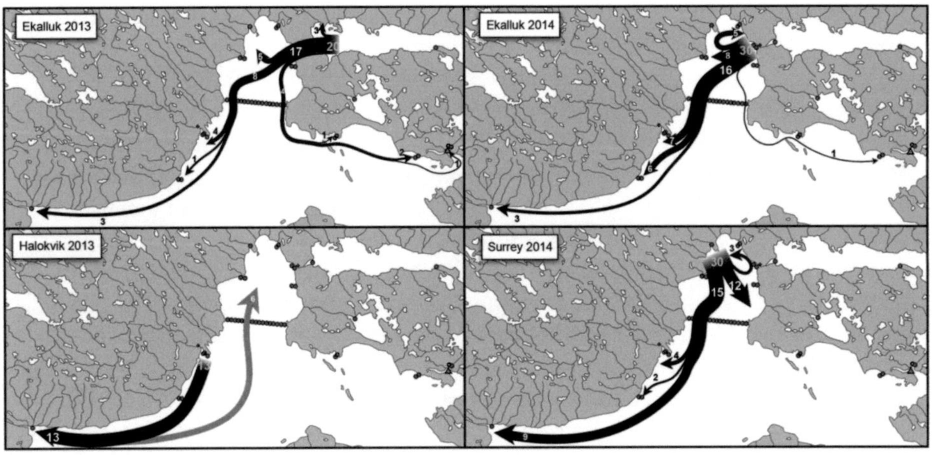

Figure 4.
General patterns of movements of Arctic charr tagged in the Cambridge Bay area, Nunavut, at three tagging locations in 2013 and 2014 and detected in summer 2014. Black arrows indicate movement from tagging location for fish tagged in 2014 or from fresh water for returning migrants tagged in 2013. The thickness of the arrows is proportional to the number of individuals observed performing a specific movement pattern (also indicated by numbers). From Howland et al. [12].

a genetic assignment-based approach combined with otolith microchemistry [72] also documented anadromous migrations in Lake trout and also noted substantial inter-lake movements in this species in the Husky Lake drainage basin, Northwest Territories, where freshwater resident, semi-anadromous, and brackish-water resident lake trout life history types are documented. Kissinger et al. [73] documented that anadromous migrations in this species can sometimes be unexpectedly longer than previously assumed.

4.2 Marine fish

4.2.1 Greenland halibut (Reinhardtius hippoglossoides)

Greenland halibut has a circumpolar distribution in the Northern Hemisphere in the North Atlantic, North Pacific, and Arctic oceans. It is an important commercial species supporting fisheries in these zones. Floy tagging between 1994 and 2000 indicated that Greenland halibut located in the northern winter fishing grounds were resident, while fish tagged near the mouth of Cumberland Sound were migratory to offshore waters of Baffin Bay and Davis Strait [19]. Western Greenland fiords have similar inshore populations of Greenland halibut that are sink populations with origins offshore but settling into the inshore [20]. It is likely that the same circumstance exists in Cumberland Sound. The existing inshore allocation or total allowable catch (TAC) of Greenland halibut was assigned to a new management area that encompasses northern Cumberland Sound. Subsequently, the question arose whether the inshore stock in Cumberland Sound was distinct from the offshore. Through acoustic telemetry monitoring of fishes at depths between 400 and 1200 m, combined with environmental and fishery data, Hussey et al. [13] examined the movement patterns of Greenland halibut in Cumberland Sound, Nunavut. They noted biotic and abiotic factors that were driving fish movements. Greenland halibut undertook clear seasonal movements between the southern and northern regions of the sound driven by temperature, dissolved oxygen, and sea ice cover, with most tagged fish using the entire sound over the course of the year.

Barkley et al. [74] used acoustic transmitters to track Greenland halibut in Scott Inlet, coastal Baffin Island, Canada, over a 1-year period. Their aim was to determine if fish could be vulnerable to both the inshore and offshore fisheries in the area.

Barkley et al. [74] described a dual pattern of movement in Greenland halibut. Most fish moved between the inshore in the summer and offshore during the winter months. A few fish moved offshore with the others but returned inshore during the main winter months. Greenland halibut seemed to avoid ice-cover in the inshore and likely moved offshore as the sea-ice formed in November.

It was thought that Greenland halibut remained within the coastal environment of Baffin Bay during the year. The recent data show that this is not the case in all areas. In the coastal regions such as Scott Inlet, inshore-offshore connectivity occurs. This implies that many of the Halibut are from a single population and vulnerable to harvest in both the inshore and the offshore fisheries. To avoid over-harvesting, fishery management must take this into account.

Further studies have noted that adult Greenland halibut may make longer migrations over wide areas for the purposes of spawning and growth (A. Fisk, University of Windsor, pers. comm.)

4.3 Marine invertebrates

4.3.1 Zooplankton

Northern shrimp are distributed patchily throughout the circumpolar region. In Canadian waters, they form important fisheries in Davis and Hudson straits and in Cumberland Sound. Webster et al. [75] noted that macrozooplankton (e.g., krill, amphipods, and jellyfish) and nekton (e.g., decapod shrimp, squid, and fish) are integral parts of pelagic ecosystems, but knowledge of their vertical distributions and migrations during winter at high latitudes is lacking. Webster et al. [75] quantified macrozooplankton and nekton distributions during the polar night in a partially ice-covered high Arctic fjord. Most nekton occurred under the 100 m thermocline both during the day and night. The nekton biomass was dominated by a varied fish community (10 species present) with shrimp and squid being the other main components. Large *Calanus* spp. copepods and gelatinous zooplankton of the macroplankton occurred all along the water column. In contrast to the nekton, the majority were above the thermocline day and night. Biomass could be predicted with a general additive model with depth, time, and moonlight. The model predicted that biomass increased with depth for both macrozooplankton (over the top 100 m) and nekton, but revealed no patterns in biomass over time.

4.3.2 Northern and striped shrimp (Pandalus borealis and P. montagui)

While the extent of large scale adult pandalid migration remains unknown due to difficulties associated with tagging and tracking of such small organisms, it is known that pelagic larvae settle in shallow waters where they reside until reaching approximately 2 years of age [76]. At that time, juvenile males migrate to deeper waters where they spend a short period of time (~1 year) before transitioning into females. Following metamorphosis, ovigerous females return to shallow waters where larvae are released. The pelagic life stage can last up to 3 months [77] and is highly influenced by oceanographic currents while settlement patterns are heavily influenced by the release location as well as the vertical migration behavior [78]. However, discrepancies still exist in our understanding of life-stage specific habitat preferences between northern and striped shrimp but there is a general agreement that *Pandalus montagui* are found in shallower (200–500 m) and cooler (−1 to 2°C) waters throughout their development [79]. Nonetheless, both species have been shown to perform diel vertical migrations to facilitate the feeding on zooplankton organisms during the night time [76, 80].

4.4 Marine mammals

4.4.1 Pinnipeds

4.4.1.1 Ringed seal (Phoca hispida)

Ringed seals are distributed throughout the Arctic Ocean. Ringed seals have a circumpolar distribution from approximately 35°N to the North Pole, occurring in all seas of the Arctic Ocean. They are hunted for food by indigenous peoples of the Canadian north and are critical to food security.

Ringed seals travel great distances during their life. Harwood et al. [81] deployed satellite-linked time-depth recorders on 17 Ringed seals in early summer in 1999, 2000, and 2010, near the Inuvialuit community of Ulukhaktok, Northwest Territories, Canada. During the open water period, mature and immature Ringed seals moved large distances (mean distance travelled = 5844 km; range of distance travelled = 1232–9473 km). Ringed Seals maintain big home ranges averaging 122,854 km^2 for immature, 76,658 and 21,649 km^2 for mature females and males, respectively, while mature seals spent the bulk of their time in a foraging/resident mode (69.5%) but also spent a lot of time moving (22.8%). Ringed seals spent three-quarters of their time in either Prince Albert Sound or eastern Amundsen Gulf. At some point, most Ringed seals did move outside the area into Prince of Wales Strait, Viscount Melville Sound, Minto Inlet, and western Amundsen Gulf. Immatures spent more time moving (36.8%) and less time foraging (51.4%) than adults. Most Ringed seals wintered in either Prince Albert Sound or Amundsen Gulf.

4.4.1.2 Walrus (Odobenus rosmarus)

Atlantic Walruses range across the Canadian Arctic, to Greenland, Svalbard, and the western part of Arctic Russia. They are occasionally harvested by Inuit for food and the tusks.

Two subspecies of Walrus occur in the Canadian Arctic: the Atlantic Walrus (*Odobenus rosmarus rosmarus*) and the Pacific Walrus (*O. r. divergens*), although the latter ranges into Canadian waters from Alaska only occasionally. The Atlantic Walrus is widely distributed throughout the eastern Canadian Arctic, where the high Arctic population occupies the waterways of the central Archipelago and northern Baffin Bay as far east as Greenland, and the central/low Arctic population occurs in northern Hudson Bay, Foxe Basin, Hudson Strait, and Davis Strait. The degree to which these populations undertake seasonal movements or migrations is not well understood, although studies in recent years have indicated both localized movements and large-scale migrations in response to seasonal variation in ice conditions [82, 83].

Walrus require large areas of open water overlying shallow (<80 m) bivalve beds, their preferred prey, with nearby ice or land for hauling out [84]. In areas where polynyas and dynamic leads in pack ice persist throughout winter, Walrus occur year-round and undertake only localized movements between wintering areas and onshore haul-out sites during summer. Walrus occur year-round in Foxe Basin, northern Hudson Bay, and western Hudson Strait [85, 86]. Although Walrus move seasonally within Foxe Basin, there is no evidence of concerted movements between Foxe Basin and Hudson Strait [87]. There is similarly no evidence of seasonal movements into or out of southeastern Hudson Bay, where Walrus move from the floe edge in winter to terrestrial haul-out sites in summer [88]. There is, however, evidence of seasonal movements between Hudson Bay and Hudson Strait, with general westward movement in summer, and return movements north and eastward in fall [89].

Historically, Walrus were known to migrate northward along West Greenland in spring and to return southward along the east coast of Baffin Island in fall (Freuchen 1921 and Vibe 1950, as cited in [89]). Although no longer observed on that scale, recent studies have confirmed seasonal Walrus migrations between Greenland and Canada. 50 Walruses from the high Arctic population, instrumented with satellite transmitters, departed their feeding banks along the Greenland coast in June–July at the onset of ice melt, swimming west into the Canadian Arctic Archipelago [83]. Tags on three of the animals transmitted long enough to document their return migration from Ellesmere Island to their original tagging locations off the Greenland coast in October [83]. Similarly, Dietz et al. [82] showed that eight of 23 Walruses from the central/low Arctic population satellite tagged at their winter grounds off central West Greenland migrated to southeast Baffin Island during April–May. Individual Walruses took on average 7 days to make the crossing, and generally followed a similar 400-km long migration path over the shallowest and narrowest part of Davis Strait [82]. As with Walrus migrations further north, the westward spring migration coincided with the seasonal retreat of the pack ice edge. Satellite transmissions did not last long enough to document a return migration, although an animal flipper-tagged off southeast Baffin Island and subsequently shot 2 years later off West Greenland provides evidence of such [82]. Telemetry results are also supported by genetics analysis that showed no differences between Walruses from West Greenland and southeast Baffin Island [90]. The seasonal movement of this species between Greenland and Canada is relevant to Walrus hunt management in both countries.

4.4.2 Cetaceans

4.4.2.1 Beluga (Delphinapterus leucas)

Beluga whale are distributed through out the Arctic Ocean and are one of two species adapted to living with pack ice. They are hunted for food by Inuit in Canada. Most Arctic populations are healthy but the population in Cumberland Sound is of concern.

Belugas take part in a yearly cycle of migration from their summering grounds to their wintering grounds. Belugas tend to frequent river estuaries and coastal water during the summer. Belugas spend the winter next to the ice edge, and in an area of open water [91]. Then, Beluga migrate back to their summering grounds. The migrations vary widely in the distance travelled but the same seasonal pattern is found in all the Arctic Beluga populations.

Beluga of the Cumberland Sound population stay within the Cumberland Sound area all year round. Their summer distribution is restricted fjords and more than half the population can be found in August in a small area of less than 150 km^2 [92]. They migrate from fjords to the open water at the mouth of Cumberland Sound in the winter [93], a migration that is only a couple hundred kilometers.

The Eastern Beaufort Sea (EBS) beluga population has a much larger summer range with the core area covering more than 50,000 km^2 [94]. Sexual segregation has been reported during the summer. Males tended to have a large home range than females [94] and they also selected areas with higher ice concentrations during July-August [95]. Females preferred areas close to the shore (<200 km) [95]. Belugas of the EBS population tended to leave their summering ground in September [94] for their fall migration westward to the Bering Sea [96]. This migration was more than 3000 km long [96].

There are two main recognized beluga populations in Hudson Bay: the Western Hudson Bay (WHB) and the Eastern Hudson Bay (EHB) [97]. Belugas in Hudson Bay spent their summer in the estuaries and river mouths. Belugas of the WHB population started their migration around mid-October [98]. EHB belugas migrated in groups of related individuals and their migration route seemed to be learned and shared by related individuals [99]. As they migrated in the spring and fall along Hudson Strait [100], both WHB and EHB belugas were susceptible to harvest by coastal aboriginal communities along their migration route [101].

4.4.2.2 Narwhal (Monodon monoceros)

Narwhal are found in Canadian, Greenlandic, and Russian Arctic waters. They are the other species adapted to exist under ice packs. They are hunted for food and their tusks by the Canadian Inuit. A tusk can be worth up to $10,000CAN.

There are five main Narwhal populations in the world: the Northern Hudson Bay, the Baffin Bay, the East Greenland, the Northeast Greenland, and the Svalbard-Russia populations. The Baffin Bay population is the largest with more than 150,000 individuals. Narwhals summer in known aggregations in bays and fjords in the high Arctic [102]. Although the location and occurrence of these summer aggregations are predictable, a proportion of narwhals seem to mix between summer aggregations [103].

Information about Narwhal migration comes from satellite telemetry and local knowledge. Narwhals started their fall migration before the ice started to form. During this critical period, they were susceptible to being trapped in ice if sudden changes in weather and wind conditions precipitated the formation of ice [104]. Narwhals from the Baffin Bay population overwintered in Baffin Bay and Davis Strait, in extreme ice coverage [105]. The total length of the migration route of Narwhal from summering to wintering ground could be more than 1000 km. Narwhals from the Northern Hudson Bay population undertook a migration of similar length between Repulse Bay, in the North West part of Hudson Bay in Canada, to the Labrador Sea [106]. Spring migrations back to the summering ground were often lead by groups of males [107]. Limited information from satellite telemetry suggests that some Narwhal returned back to the same summering ground [108]. However, satellite telemetry data also showed that Narwhal can change between summering ground [103].

4.4.2.3 Bowhead (Balaena mysticetus)

Bowhead whale live entirely in the Arctic and Sub-Arctic waters of the northern hemisphere. They complete their entire life cycle in the north. Hunting for bowhead whales commercially nearly extirpated populations but now is banned. Inuit in Nunavut are allocated up to three whales annually for subsistence harvest by communities. Hunts occur in different communities each time. Hunt planning and permits take several years and the hunt is closely monitored.

There are four bowhead whale populations around the circumpolar Arctic and the two largest spend the majority of their time in Canada—the Bering-Chukchi-Beaufort (BCB) Sea and Eastern Canada-West Greenland (ECWG) populations. Both populations migrate great distances, as would be expected for large-bodied marine mammals [109] and can cross paths within the Canadian Archipelago [95]. The main location of seasonally rich food supplies occurs in the polar summer and is spatially and temporally separated from environments used for mating, calving, and lactation [110]. There is an increased understanding of Bowhead migratory behavior because of the use of new satellite telemetry techniques to track marine mammal migration [111, 112]. In some cases, these migrations follow predictable routes but age, sex, and reproductive spatial segregation occurs [113].

The BCB population travels more directly from their winter range because the greater amount of clear water and a small amount of pack ice allow easier movements [114]. After the winter, the spring migration is to the north and east into the Beaufort Sea [115]. During summer, they live principally in the Canadian part of the Beaufort in the Amundsen Gulf and Viscount-Melville areas [116]. Finally, the fall migration takes the bowheads through the Alaskan Beaufort and Chukchi Seas, and then through the Bering Strait into the Bering Sea. Within Canadian waters, the total range of the Bering-Chukchi-Beaufort Sea population is approximately 207,000 km^2.

In contrast, the ECWG population circumnavigates Baffin Island with the extent of occurrence of roughly 1 million km^2 [58]. Wintering occurs in areas with unconsolidated pack ice, particularly in Hudson Strait and along the southeastern Baffin Bay coastline. A segment of the ECWG Bowhead whale population moves east to the West Greenland coast (Disko Bay) in late winter-early spring, likely mature adults involved in mating. Another segment of the population consisting mostly of females with calves and juveniles moves west and north into Foxe Basin [117]. In summer, the Greenland portion travels north and west into western Baffin Bay, whereas the portion in Foxe Basin travels through Fury and Hecla Strait into the Canadian High Arctic, particularly Gulf of Boothia and Prince Regent Inlet, where they are met by the Greenland portion of the population [118]. The fall migration occurs over 2–3 months starting in late August/September with whales either moving back through Fury and Hecla Strait into northern Hudson Bay or along the east coast of Baffin Island to winter areas in the south [119]. Peak feeding is thought to occur during the fall migration and areas such as Isabella Bay along the east coast of Baffin Island are considered key foraging habitat [120]. Bowhead whales appear to follow the waxing and waning of seasonal sea ice presumably because these areas provide access to food and shelter/protection from killer whale predation [111]. Large adults are likely able to defend against killer whales with access to coastlines and shallow water and therefore use open water areas along the Greenland and Baffin Island coastlines during summer and autumn [121]. Whales that are smaller and more at risk of killer whale predation stay.

5. Conclusions

The Canadian Arctic is a region of great ecological variability. The species that have evolved in this area make use of the variable opportunities presented by undertaking extensive migrations. These migratory activities have consequences to the population structure and to the availability of the biota to humans. For example, anadromous fishes perform precise long distance migrations for the purposes of reproduction travelling from marine feeding areas to freshwater spawning areas. They subsequently have a complex population structure with many unique populations or stocks. Inter-population genetic variability is high. It is likely that the genetic adaptability of the salmonidae is why they dominate in northern zones. Their travels also make them most available to human predators and the accessibility to harvest is frequently the reason for the location of communities in the Canadian Arctic. Accordingly, resource management must be stock-by-stock on relatively small geographical scales. In contrast, the main harvested marine fish species, Greenland halibut, appears to have little population structuring over vast areas. The migrations do not segregate them into distinct stocks. Probably, the use of planktonic larvae is related to this and that the profundal zones where they live as adults are not especially variable. Finally, marine mammals show a population structure but not to the same level as fishes. Their migratory abilities are the most pronounced but the segregation between them is more likely a result of learning than instinct.

Migratory patterns are highly variable among harvested aquatic organisms in the Arctic but the understanding of the patterns of migration is critical for successful harvesting and management. The patterns of migration in freshwater and anadromous fishes have often determined the location of indigenous communities, which may have started as simple fishing camps. It is likely that the patterns of migration in marine mammals also have had a large influence on northern development in Canada. Knowledge of the precise time and location of migrations was essential for survival in a harsh and unforgiving region. In contrast, the migrations and segregation of offshore marine resources are relatively imprecise. Development of fisheries has come with the induction of "western technologies" of large industrial scale ships and nets. The knowledge of migratory patterns is relevant to designating management zones but not to survival. The migratory patterns of aquatic organisms have shaped northern culture, communities, and the way of life for aboriginal populations.

Author details

Ross F. Tallman[1*], Steven H. Ferguson[1], Les N. Harris[1], Kevin J. Hedges[1], Kimberly L. Howland[1], Nigel E. Hussey[2], Marianne Marcoux[1], Cory J.D. Matthews[1], Zoya A. Martin[3] and Jean-Sebastien Moore[4]

1 Fisheries and Oceans Canada Arctic Region, Winnipeg, Manitoba, Canada

2 Department of Biological Sciences, University of Windsor, Windsor, Ontario, Canada

3 Fisheries and Oceans Canada Arctic Region, Iqaluit, Nunavut, Canada

4 Département de Biologie, Université de Laval, Quebec City, Quebec, Canada

*Address all correspondence to: ross.tallman@dfo-mpo.gc.ca

IntechOpen

© 2019 The Author(s). Licensee IntechOpen. This chapter is distributed under the terms of the Creative Commons Attribution License (http://creativecommons.org/licenses/by/3.0), which permits unrestricted use, distribution, and reproduction in any medium, provided the original work is properly cited.

References

[1] Binder TR, Cooke SJ, Hinch SG. The biology of fish migration. In: Farrell AP, editor. Encyclopedia of Fish Physiology: From Genome to Environment. Vol. 3. San Diego: Academic Press; 2011. pp. 1921-1927

[2] Tallman RF, Saurette F, Thera T. Migration and life history variation in Arctic charr, *Salvelinus alpinus*. Ecoscience. 1996;**3**(1):33-41

[3] Power G. A review of fish ecology in Arctic North America. In: Reynolds J, editor. Fish Ecology in Arctic North America Bethesda: American Fisheries Society Symposium. pp. 13-39

[4] Stearns SC. The Evolution of Life Histories. Oxford: Oxford University Press; 1992. 249 p

[5] Myers GS. Usage of anadromous, catadromous and allied terms for migratory fishes. Copeia. 1949;**1949**:89-97

[6] Northcote TG. Migratory strategies and production in freshwater fishes. In: Gerking SD, editor. Ecology of Freshwater Fish Production. Oxford: Blackwell: Sci. Publishers; 1978. pp. 326-329

[7] Gross MR. Evolution of diadromy in fishes. American Fisheries Society Symposium. 1987;**1**:14-25

[8] Hamilton WD, May RM. Dispersal in stable habitats. Nature. 1977;**269**(5629):578

[9] Christiansen JS, Mecklenburg CW, Karamushko OV. Arctic marine fishes and their fisheries in light of global change. Global Change Biology. 2014;**20**(2):352-359

[10] MacNeil MA, Graham NA, Cinner JE, Dulvy NK, Loring PA, Jennings S, et al. Transitional states in marine fisheries: Adapting to predicted global change. Philosophical Transactions of the Royal Society, B: Biological Sciences. 2010;**365**(1558):3753-3763

[11] Tallman RF, Thera TM. A life-cycle based fisheries model for broad whitefish, *Coregonus nasus*, of the lower Mackenzie River: A solution to management of a trans-boundary migratory species in the Canadian Arctic. Journal of Fish Biology. 2006;**69**:244-244

[12] Howland K, Mochnacz N, Gallagher C, Tallman R, Ghamry H, Roux M-J, et al. Developing strategies for improved assessment and ecosystem-based management of Canadian Northern Dolly Varden. In: Kruse GH, Browman HI, Cochrane KL, Evans D, Jamieson GS, Livingston PA, Woodby D, Zhang CI, editors. Global Progress in Ecosystem-Based Fisheries Management. Alaska Sea Grant: University of Alaska Fairbanks; 2012. pp. 269-188

[13] Hussey NE, Hedges KJ, Barkley AN, Treble MA, Peklova I, Webber DM, et al. Movements of a deep-water fish: Establishing marine fisheries management boundaries in coastal Arctic waters. Ecological Applications. 2017;**27**(3):687-704

[14] Gregg JL, Anderl DM, Kimura DK. Improving the precision of otolith-based age estimates for Greenland halibut (*Reinhardtius hippoglossoides*) with preparation methods adapted for fragile sagittae. Fishery Bulletin. 2006;**104**(4):643

[15] Treble MA, Campana SE, Wastle RJ, Jones CM, Boje J. Growth analysis and age validation of a Deepwater Arctic fish, the Greenland halibut (*Reinhardtius hippoglossoides*). Canadian Journal of Fisheries and Aquatic Sciences. 2008;**65**(6):1047-1059

[16] Scott WB, Scott MG. Atlantic fishes of canada. Canadian Bulletin of Fisheries and Aquatic Sciences. 0706-6503219. 1988. 791 p

[17] Bowering WR, Chumakov AK. Distribution and relative abundance of Greenland halibut (*Reinhardtius hippoglossoides* (Walbaum)) in the Canadian Northwest Atlantic from Davis Strait to the northern Grand Bank. Fisheries Research. 1989;**7**(4):301-327

[18] Bowering WR, Brodie WB. Greenland halibut (*Reinhardtius hippoglossoides*). A review of the dynamics of its distribution and fisheries off eastern Canada and Greenland. In: Deep-Water Fisheries of the North Atlantic Oceanic Slope. Dordrecht: Springer; 1995. pp. 113-160

[19] Treble MA, Jørgensen OA. Summary of results for Greenland halibut from trawl surveys conducted in NAFO subareas 0 and 1 from 61 N to 74 N in 2001. NAFO SCR Doc, 2; 2002. p. 60

[20] Treble MA, Jørgensen OA. Implementation of advice for data limited stocks: A survey approach for Greenland halibut in SA 0+. Northwest Atlantic Fishery Organization - SCR Documents. 2015. 15/035 5p

[21] Boje J. Intermingling and seasonal migrations of Greenland halibut (*Reinhardtius hippoglossoides*) populations determined from tagging studies. Fishery Bulletin. 2002;**100**(3):414-422

[22] Howland KL. Migration patterns of freshwater and anadromous inconnu (*Stenodus leucichthys*) within the Mackenzie River system. M.Sc. Dissertation University of Alberta. 1997. 97 p

[23] Howland KL, Tallman RF, Tonn WM. Migration patterns of freshwater and anadromous inconnu in the Mackenzie River system. Transactions of the American Fisheries Society. 2000;**129**(1):41-59

[24] Howland KL, Tonn WM, Babaluk JA, Tallman RF. Identification of freshwater and anadromous inconnu in the Mackenzie River system by analysis of otolith strontium. Transactions of the American Fisheries Society. 2001;**130**(5):725-741

[25] Stephenson SA, Burrows JA, Babaluk JA. Long-distance migrations by inconnu (*Stenodus leucichthys*) in the Mackenzie River system. Arctic. 2005;**58**:21-25

[26] Harris LN, Loewen LT, Reist JD, Tallman RF, Babaluk JA, Halden NM. Migratory variation of Mackenzie River system broad whitefish: Insights from otolith strontium distributions. Transactions of the American Fisheries Society. 2012;**141**:1574-1585

[27] Harris LN, Taylor EB, Tallman RF, Reist JD. Gene flow and effective population size in two life-history types of broad whitefish, *Coregonus nasus*, from the Canadian Arctic. Journal of Fish Biology. 2012;**81**:288-307

[28] Tallman RF, Abrahams MV, Chudobiak DH. Migration and life history alternatives in a high latitude species, the broad whitefish, *Coregonus nasus* Pallas. Ecology of Freshwater Fish. 2002;**11**(2):101-111

[29] Chang-Kue KTJ, Jessop EF. Broad whitefish radio-tagging studies in the lower Mackenzie River and adjacent coastal region, 1982-1993. In: The Proceedings of the Broad Whitefish Workshop: The Biology, Traditional Knowledge and Scientific Management of Broad Whitefish (Coregonus nasus (Pallas)) in the Lower Mackenzie River. Ross F. Tallman, James D. Reist editors. Canadian technical report of fisheries and aquatic sciences; 1997;**2193**:117-146

[30] Harris LN, Taylor EB. Genetic population structure of broad whitefish, *Coregonus nasus*, from the Mackenzie River, Northwest Territories: Implications for subsistence fishery management. Canadian Journal of Fisheries and Aquatic Sciences. 2010;**67**(6):905-918

[31] Chudobiak DH, Tallman RF, Abrahams MV. Variation in the morphology of two populations of Arctic broad whitefish, *Coregonus nasus* (Pallas), in the Mackenzie River, Northwest Territories, Canada. Ergebnisse der Limnologie. 2002;**57**:291-305

[32] Millar N, Harris L, Howland K. Seasonal migration of broad whitefish (*Coregonus nasus* (Pallas)) in an Arctic lake. Advances in Limnology. 2013;**64**:91-107

[33] Reist JD, Bond WA. Life history characteristics of migratory coregonids of the lower Mackenzie River, Northwest Territories, Canada. Finnish Fisheries Research. 1988;**9**:133-144

[34] Fechhelm RG, Bryan JD, Griffiths WB, Wilson WJ, Gallaway BJ. Effect of coastal winds on the summer dispersal of young least cisco (*Coregonus sardinella*) from the Colville River to Prudhoe Bay, Alaska: A simulation model. Canadian Journal of Fisheries and Aquatic Sciences. 1994;**51**(4):890-899

[35] Tallman RF, Howland KL. Seasonal migration patterns of lower Mackenzie River coregonids. Advances in Limnology. 2008;**63**:133-146

[36] Griffiths WB, Gallaway BJ, Gazey WJ, Dillinger RE Jr. Growth and condition of Arctic cisco and Broad whitefish as indicators of causeway-induced effects in the Prudhoe Bay region, Alaska. Transactions of the American Fisheries Society. 1992;**121**(5):557-577

[37] Fechhelm RG, Griffiths WB. Effect of wind on the recruitment of Canadian Arctic cisco (*Coregonus autumnalis*) into the central Alaskan Beaufort Sea. Canadian Journal of Fisheries and Aquatic Sciences. 1990;**47**(11):2164-2171

[38] Dillinger RE Jr, Birt TP, Green JM. Arctic cisco, *Coregonus autumnalis*, distribution, migration and spawning in the Mackenzie River. Canadian Field-Naturalist. 1992;**106**(2):175-180

[39] Zimmerman CE, Ramey AM, Turner SM, Mueter FJ, Murphy SM, Nielsen JL. Genetics, recruitment, and migration patterns of Arctic cisco (*Coregonus autumnalis*) in the Colville River, Alaska, and Mackenzie River, Canada. Polar Biology. 2013;**36**(11):1543-1555

[40] Bond WA, Erickson RN. Coastal migrations of Arctic ciscoes in the eastern Beaufort Sea. In: Fish ecology in Arctic North America. American Fisheries Society Symposium. No. 19. 1997. pp. 155-164

[41] Lawrence MJ, Lacho G, Davies S.: A survey of the coastal fishes of the Southeastern Beaufort Sea. Canadian Technical Report Fisheries and Aquatic Sciences I 220. 1984

[42] Hatfield CT, Stein JN, Falk MR, Jessop CS.: Fish Resources of the Mackenzie River Valley. Interim Report I. Vol. I. Winnipeg, MB: Department of the Environment, Fisheries Service;. 1972. 247 p

[43] Lange M, Tallman RF. Section IV-Biology-Life history variation in two populations of migratory lake whitefish (*Coregonus clupeaformis*) in the western Northwest Territories, Canada (with 6 figures). Ergebnisse der Limnologie. 2002;**57**:507-516

[44] Kowalchuk MW, Sawatzky CD, Reist JD. A review of the taxonomic

structure within Dolly Varden, *Salvelinus malma* (Walbaum 1792), of North America. Canadian Science Advisory Secretariat Research Document 013; 2010. pp. i-vi, 1-16

[45] Harris LN, Bajno R, Gallagher CP, Koizumi I, Johnson LK, Howland KL, et al. Life-history characteristics and landscape attributes as drivers of genetic variation, gene flow, and fine-scale population structure in northern Dolly Varden (*Salvelinus malma malma*) in Canada. Canadian Journal of Fisheries and Aquatic Sciences. 2015;**72**(10):1477-1493

[46] Gallagher C, Morisson C, Lea E, Halden N, Howland K. Growth and reproductive characteristics of rarely observed female resident Dolly Varden (*Salvelinus malma malma*) in North America. Hydrobiologia. 2019 (in press)

[47] Harwood LA, Sandstrom S, Linn E. Status of anadromous Dolly Varden (*Salvelinus malma*) of the Rat River, Northwest Territories, as assessed through sampling of the subsistence fishery (1995-2007). Canadian Manuscript Report Fisheries and Aquatic Science. 2891. 2009. vii + 52 p

[48] Courtney MB, Scanlon B, Brown RJ, Rikardsen AH, Gallagher CP, Seitz AC. Offshore Ocean dispersal of adult Dolly Varden *Salvelinus malma* in the Beaufort Sea. Polar Biology. 2018;**41**:817-825

[49] Morrison C. Life history strategies of northern form Dolly Varden (*Salvelinus malma* malma) in the western Canadian Arctic [MSc. Thesis University of Alberta]. 2017. 114 p

[50] Armstrong RH. Morrow JE. The dolly varden. Charrs: salmonid fishes of the genus Salvelinus. 1980. pp. 99-140

[51] McCart PJ. A review of the systematics and ecology of Arctic charr, *Salvelinus alpinus*, in the western Arctic. Canadian Technical Report of Fisheries and Aquatic Sciences 935. 1980

[52] Stewart DB, Mochnacz NJ, Reist JD, Carmichael TJ, Sawatzky CD. Fish life history and habitat use in the Northwest Territories: Dolly Varden (*Salvelinus malma*). Canadian Manuscript Report Fisheries and Aquatic Science. 2915. 2010. vi + 63 p

[53] Craig PC. Ecological studies of anadromous populations of Arctic charr in the canning river drainage and adjacent coastal waters of the Beaufort Sea, Alaska. In: McCart PJ, editor. Fisheries investigatons along the north slope and Beaufort sea coast in Alaska with emphasis on Arctic charr. Calgary, Alberta: Canadian Arctic Gas Study Ltd.; 1977. pp. 1-116

[54] Underwood TJ, Millard MJ, Thorpe LA. Relative abundance, length frequency, age, and maturity of Dolly Varden in nearshore waters of the Arctic National Wildlife Refuge, Alaska. Transactions of the American Fisheries Society. 1996;**125**(5):719-728

[55] DeCicco A. Movements and spawning of adult Dolly Varden charr (*S. malma*) in Chukchi Sea drainages of northwestern Alaska: Evidence for summer and fall spawning populations. In: Kawanabe H, Yamazaki F, Noakes DLG editors. Biology of Charrs and Masu Salmon. In: Proceedings of the International Symposium on Charrs and Masu Salmon. Physiology and Ecology Japan, Special. 1989;**1**:229-238

[56] Sandstrom SJ, Chetkiewicz CB, Harwood LA. Overwintering habitat of juvenile Dolly Varden (*Salvelinus malma*) (W.) in the Rat River, NT, as determined by radio telemetry. Canadian Science Advisory Secretariat 2001/092. 2001

[57] Glova G, McCart PJ. Life history of Arctic charr (*Salvelinus alpinus* L.) in

the Firth River system, Yukon Territory. In: McCart PJ, editor. Life Histories of Anadromous and Freshwater Fish in the Western Arctic. Calgary, Alberta: Canadian Arctic Gas Study Ltd; 1980

[58] Sandstrom SJ, Harwood LA. Studies of anadromous Dolly Varden (*Salvelinus malma*) (W.), of the Big Fish River, NT, Canada 1972-1994. Canadian Manuscript Report of Fisheries and Aquatic Sciences 2603. 2002. i-vi, 1-31 p

[59] Gallagher CP, Howland KL, Sandstrom SJ, Halden NM. Migration tactics affect spawning frequency in an iteroparous salmonid (*Salvelinus malma*) from the Arctic. PLoS One. 2018;**13**(12):e0210202. https://doi.org/10.1371/journal.pone.0210202

[60] DeCicco AL. Long-distance movements of anadromous Dolly Varden between Alaska and the U.S.S.R. Arctic. 1992;**45**(2):120-123

[61] Moore J-S, Harris LN, Tallman RF, Taylor EB. The interplay between dispersal and gene flow in anadromous Arctic charr (*Salvelinus alpinus*): Implications for potential for local adaptation. Canadian Journal of Fisheries and Aquatic Sciences. 2013;**70**:1327-1338

[62] Harris LN, Moore J-S, Galpern P, Tallman RF, Taylor EB. Geographic influences on fine-scale, hierarchical population structure in northern Canadian populations of anadromous Arctic charr (*Salvelinus alpinus*). Environmental Biology of Fishes. 2014;**97**:1233-1252

[63] Moore J-S, Harris LN, Le Luyer J, Sutherland B, Rougemont Q, Tallman RT, et al. Migration harshness drives habitat choice and local adaptation in anadromous Arctic charr: Evidence from integrating population genomics and acoustic telemetry. Molecular Ecology. 2017;**26**(24):6784-6800

[64] Harris LN, Moore J-S, Bajno R, Tallman RF. Genetic population structure of anadromous Arctic charr (*Salvelinus alpinus*) from the Cambridge Bay Region, NU: Potential implications for the management of Canada's largest Arctic charr commercial fishery. North American Journal of Fisheries Management. 2016;**63**(2):1473-1488

[65] Johnson L. The Arctic charr, *Salvelinus alpinus*. In: Balon EK, editor. Charrs: Salmonid Fishes of the Genus *Salvelinus*. The Hague, The Netherlands: Dr. W. Junk BV; 1980. pp. 15-98

[66] Dempson JB, Kristofferson AH. Spatial and temporal aspects of the ocean migration of anadromous Arctic charr. In: Common Strategies of Anadromous and Catadromous Fishes. Dadswell MJ, Klauda RJ, Moffitt CM, Saunders RL, editors. American Fisheries Society; 1987. pp. 340-357

[67] Spares AD, Stokesbury MJW, O'Dor RK, Dick TA. Temperature, salinity and prey availability shape the marine migration of Arctic charr, *Salvelinus alpinus*, in a macrotidal estuary. Marine Biology. 2012;**159**(8):1633-1646. DOI: 10.1007/s00227-012-1949-y

[68] Spares AD, Stokesbury M, Dadswell MJ, O'Dor RK, Dick TA. Residency and movement patterns of Arctic charr *Salvelinus alpinus* relative to major estuaries. Journal of Fish Biology. 2015;**86**:1754-1780. DOI: 10.1111/jfb.12683

[69] Moore JS, Harris LN, Kessel ST, Bernatchez L, Tallman RF, Fisk AT. Preference for nearshore and estuarine habitats in anadromous Arctic charr (*Salvelinus alpinus*) from the Canadian high Arctic (Victoria Island, Nunavut) revealed by acoustic telemetry. Canadian Journal of Fisheries and Aquatic Sciences. 2016;**73**(9):1434-1445

[70] Harris LN, Chavarie L, Bajno R, Howland KL, Wiley SH, Tonn WM, et al. Evolution and origin of sympatric shallow-water morphotypes of Lake Trout, Salvelinus namaycush, in Canada's Great Bear Lake. Heredity. 2015;**114**(1):94

[71] Swanson HK, Kidd KA, Babaluk JA, Wastle RJ, Yang PP, Halden NM, et al. Anadromy in Arctic populations of lake trout (*Salvelinus namaycush*): Otolith microchemistry, stable isotopes, and comparisons with Arctic charr (*Salvelinus alpinus*). Canadian Journal of Fisheries and Aquatic Sciences. 2010;**67**:842-853

[72] Harris LN, Moore JS, McDermid CG, Swanson HK. Long-distance anadromous migration in a fresh water specialist: The lake trout (*Salvelinus namaycush*). The Canadian Field-Naturalist. 2014;**128**(3):260-264

[73] Kissinger BC, Bystriansky J, Czehryn N, Enders EC, Treberg J, Reist JD, et al. Environment-phenotype interactions: Influences of brackish-water rearing on lake trout (*Salvelinus namaycush*) physiology. Environmental Biology of Fishes. 2017;**100**(7):797-814

[74] Barkley AN, Fisk AT, Hedges KJ, Treble MA, Hussey NE. Transient movements of a deep-water flatfish in coastal waters: Implications of inshore-offshore connectivity for fisheries management. Journal of Applied Ecology. 2018;**55**(3):1071-1081

[75] Webster CN, Varpe Ø, Falk-Petersen S, Berge J, Stübner E, Brierley AS. Moonlit swimming: Vertical distributions of macrozooplankton and nekton during the polar night. Polar Biology. 2015;**38**(1):75-85

[76] Shumway SE, Perkins HC, Schick DF, Stickney AP. Synopsis of biological data of the Pink Shrimp *Pandalus borealis* (KrØyer 1838). FAO Fisheries Synposis No. 144; NOAA Technical Report of the National Marine Fisheries Service. 1985. 57 p

[77] Ouellet P, Chabot D. Rearing *Pandalus borealis* (KrØyer) larvae in the laboratory: I. Development and growth at three temperatures. Marine Biology. 2005;**147**:869-880

[78] Le Corre N, Pepin P, Han G, Ma Z, Snelgrove PVR. Assessing the connectivity patterns among management units of the Newfoundland and Labrador shrimp populations. Fisheries Oceanography. 2019;**28**:183-202

[79] DFO. Assessment of Northern Shrimp, *Pandalus borealis*, and Striped Shrimp, *Pandalus montagui*, in the Eastern and Western Assessment Zones, DFO Canadian Science Advisory Secretariat Science Advisory Report 2017/010. 2017

[80] Crawford RE, Hudon H, Parsons DG. An acoustic study of shrimp (*Pandalus montagui*) distribution near Resolution Island (eastern Hudson Strait). Canadian Journal of Fisheries and Aquatic Sciences. 1992;**49**:842-856

[81] Harwood LA, Smith TG, Auld J, Melling H, Yurkowski DJ. Seasonal movements and diving of ringed seals, Pusa hispida, in the Western Canadian Arctic, 1999-2001 and 2010-11. Arctic. 2015:193-209

[82] Dietz R, Born EW, Stewart REA, Heide-Jørgensen MP, Stern H, Rigét F, et al. Movements of walrus (*Odobenus rosmarus*) between central West Greenland and Southeast Baffin Island, 2005-2008. NAMMCO Scientific Publications. 2014;**9**:53-74

[83] Heide-Jørgensen MP, Flora J, Anderson AO, Stewart REA, Nielsen NH, Hansen RG. Walrus movements in Smith sound: A Canada-Greenland shared stock. Arctic. 2017;**70**:308-318

[84] Born EW, Gjertz I, Reeves RR. Population Assessent of Atlantic Walrus. Meddelelser Number. Oslo: Norsk Polarinst. 1995;**138**:100

[85] Orr JR, Rebizant T. A summary of information on the seasonal distribution and abundance of Walrus (*Odobenus rosmarus*) in the area of northern Hudson Bay and western Hudson Strait, NWT, as collected from local hunters. Canadian Data Report of Fisheries and Aquatic Sciences 624. 1987. pp. iv + 16

[86] Elliott RE, Moulton VD, Raborn SW, Davis RA. Hudson Strait marine mammal surveys, 10 March–2 April 2012. LGL Report No. TA8129-2. Prepared by LGL Limited, King City, ON for Baffinland Iron Mines Corporation; Toronto ON. 2013. 87 p

[87] Anderson LE, Garlich-Miller J. Economic analysis of the 1992 and 1993 summer Walrus hunts in northern Foxe Basin, Northwest Territories. Canadian Technical Report of Fisheries Aquatic Sciences 2011. 1994. pp. iv + 20

[88] Hammill MO, Mosnier A, Gosselin J-F, Higdon JW, Stewart DB. Doniol-Valcroze T, et al. Estimating abundance and total allowable removals for Walrus in the Hudson Bay-Davis Strait and south and east Hudson Bay stocks during September 2014. DFO Canadian Science Advisory Secetariat Research Document. 2016/036. 2016. pp. v + 37

[89] COSEWIC. COSEWIC assessment and status report on the Atlantic Walrus *Odobenus rosmarus rosmarus*, High Arctic population, Central-Low Arctic population and Nova Scotia-Newfoundland-Gulf of St. Lawrence population in Canada. Committee on the Status of Endangered Wildlife in Canada. Ottawa. 2017. pp. xxi + 89 Available from: http://www.registrelep-sararegistry.gc.ca/default.asp?lang=en&n=24F7211B-1

[90] Andersen LW, Born EW, Stewart RE, Dietz R, Doidge DW, Lanthier C. A genetic comparison of West Greenland and Baffin Island (Canada) walruses: Management implications. NAMMCO Scientific Publications. 2014;**9**:33-52

[91] Heide-Jørgensen MP, Sinding M-HS, Nielsen NH, Rosing-Asvid A, Hansen RG. Large numbers of marine mammals winter in the north water polynya. Polar Biology. 2016:1-10. DOI: 10.1007/s00300-015-1885-7

[92] Marcoux M, Young BG, Asselin NC, Watt CA, Dunn JB, Ferguson SH. Estimate of Cumberland Sound beluga (*Delphinapterus leucas*) population size from the 2014 visual and photographic aerial survey. DFO Canadian Science Advisory Secretariat Research Document 2016/037. 2016. iv + 19 p

[93] Richard P, Stewart DB. Information relevant to the identification of critical habitat for Cumberland Sound Belugas (*Delphinapterus leucas*). Canadian Science Advisory Secretariat Research Document 2008/085. 2008

[94] Hauser DD, Laidre KL, Suydam RS, Richard PR. Population-specific home ranges and migration timing of Pacific Arctic beluga whales (*Delphinapterus leucas*). Polar Biology. 2014;**37**(8):1171-1183

[95] Hauser DDW, Laidre KL, Stern HL, Moore SE, Suydam RS, Richard PR. Habitat selection by two beluga whale populations in the Chukchi and Beaufort seas. PLoS One. 2017;**12**(2):e0172755. DOI: 10.1371/journal.pone.0172755

[96] Citta JJ, Richard P, Lowry LF, O'Corry-Crowe G, Marcoux M, Suydam R, et al. Satellite telemetry reveals population specific winter ranges of beluga whales in the Bering Sea. Marine Mammal Science. 2017;**33**(1):236-250. DOI: 10.1111/mms.12357

[97] Richard PR. Stock definition of belugas and narwhals in Nunavut. DFO

Canadian Science Advisory Secretariat. Research Document 2010/022. 2010. pp. iv + 14

[98] Pirotta E, New L, Marcoux M. Modelling beluga habitat use and baseline exposure to shipping traffic to design effective protection against prospective industrialization in the Canadian Arctic. Aquatic Conservation: Marine and Freshwater Ecosystems. 2018. DOI: 10.1002/aqc.2892

[99] Colbeck GJ, Duchesne P, Postma LD, Lesage V, Hammill MO, Turgeon J. Groups of related belugas (*Delphinapterus leucas*) travel together during their seasonal migrations in and around Hudson Bay. Proceedings of the Royal Society B: Biological Sciences. 7 Feb 2013;**280**(1752). https://doi.org/10.1098/rspb.2012.2552

[100] Lewis A, Hammill M, Power M, Doidge D, Lesage V. Movement and aggregation of eastern Hudson Bay beluga whales (*Delphinapterus leucas*): A comparison of patterns found through satellite telemetry and Nunavik traditional ecological knowledge. Arctic. 2009;**62**:13-24

[101] Rioux È, Lesage V, Postma L, Pelletier É, Turgeon J, Stewart R, et al. Determining harvest composition and wintering assemblages of belugas at a contemporary ecological scale using stable isotopes and trace elements. Endangered Species Research. 2012. DOI: 10.3354/esr00445

[102] Heide-Jørgensen MP, Richard PR, Dietz R, Laidre KL. A metapopulation model for Canadian and West Greenland narwhals. Animal Conservation. 2013;**16**(3):331-343

[103] Watt CA, Orr J, LeBlanc B, Richard P, Ferguson SH. Satellite tracking of narwhals (*Monodon monoceros*) from Admiralty Inlet (2009) and Eclipse Sound (2010-2011). Research Document, Canadian Science Advisory Secretariat. 2012

[104] Laidre K, Heide-Jørgensen MP, Stern H, Richard P. Unusual narwhal sea ice entrapments and delayed autumn freeze-up trends. Polar Biology. 2012;**35**:149-154

[105] Laidre KL, Heide-Jorgensen MP. Arctic sea ice trends and narwhal vulnerability. Biological Conservation. 2005;**121**(4):509-517

[106] Westdal KH, Richard PR, Orr JR. Migration route and seasonal home range of the northern Hudson Bay narwhal (*Monodon monoceros*). In: Ferguson SH, Loseto LL, Mallory ML, editors. A Little less Arctic: Top Predators in the World's Largest Northern Inland Sea, Hudson Bay. New York: Springer Netherlands; 2010. pp. 71-91

[107] Greendale RG, Brousseau-Greendale C. Observations of marine mammals at Cape Hay, Bylot Island during the summer of 1976. Department of the Environment Fisheries and Marine Service. Technical Report No. 680. 1976. ix + 25 p

[108] Heide-Jørgensen MP, Nielsen NH, Hansen RG, Schmidt HC, Blackwell SB, Jørgensen OA. The predictable narwhal: Satellite tracking shows behavioural similarities between isolated subpopulations: Satellite tracking isolated populations of narwhals. Journal of Zoology. 2015;**297**(1):54-65. DOI: 10.1111/jzo.12257

[109] Boyd IL. Migration of marine mammals. In: Biological Resources and Migration. Berlin, Heidelberg: Springer; 2004. pp. 203-210

[110] Harwood LA, Quakenbush LT, Small RJ, George JC, Pokiak J, Pokiak C, et al. Movements and inferred foraging by bowhead whales in the Canadian Beaufort Sea during August and September, 2006-12. Arctic. 2017;**70**:161-176

[111] Ferguson SH, Dueck L, Loseto LL, Luque SP. Bowhead whale *Balaena mysticetus* seasonal selection of sea ice. Marine Ecology Progress Series. 2010;**411**:285-297

[112] Citta JJ, Okkonen SR, Quakenbush LT, Maslowski W, Osinski R, George JC, et al. Oceanographic characteristics associated with autumn movements of bowhead whales in the Chukchi Sea. Deep Sea Research Part II: Topical Studies in Oceanography. 2018;**152**:121-131

[113] Cubbage JC, Calambokidis J. Size-class segregation of bowhead whales discerned through aerial stereophotogrammetry. Marine Mammal Science. 1987;**3**(2):179-185

[114] Citta JJ, Quakenbush LT, George JC, Small RJ, Heide-Jørgensen MP, Brower H, et al. Winter movements of bowhead whales (*Balaena mysticetus*) in the Bering Sea. Arctic. 2012;**65**:13-34

[115] Quakenbush LT, Citta JJ, George JC, Small RJ, Heide-Jørgensen MP. Fall and winter movements of bowhead whales (*Balaena mysticetus*) in the Chukchi Sea and within a potential petroleum development area. Arctic. 2010;**63**:289-307

[116] Heide-Jørgensen MP, Laidre KL, Quakenbush LT, Citta JJ. The Northwest Passage opens for bowhead whales. Biology Letters. 2011;**8**(2):270-273

[117] Pomerleau C, Patterson TA, Luque S, Lesage V, Heide-Jørgensen MP, Dueck LL, et al. Bowhead whale *Balaena mysticetus* diving and movement patterns in the eastern Canadian Arctic: Implications for foraging ecology. Endangered Species Research. 2011;**15**(2):167-177

[118] Heide-Jørgensen MP, Laidre KL, Wiig Ø, Postma L, Dueck L, Bachmann L. Large-scale sexual segregation of bowhead whales. Endangered Species Research. 2010;**13**(1):73-78

[119] Chambault P, Albertsen CM, Patterson TA, Hansen RG, Tervo O, Laidre KL, et al. Sea surface temperature predicts the movements of an Arctic cetacean: The bowhead whale. Scientific Reports. 2018;**8**(1):9658

[120] Finley KJ. Isabella Bay, Baffin Island: An important historical and present-day concentration area for the endangered bowhead whale (*Balaena mysticetus*) of the eastern Canadian. Arctic. 1990;**43**:137-152

[121] Nielsen NH, Laidre K, Larsen RS, Heide-Jørgensen MP. Identification of potential foraging areas for bowhead whales in Baffin Bay and adjacent waters. Arctic. 2015;**68**:169-179

[122] AMAP Assessment Report: Arctic Monitoring and Assessment Programme (AMAP). Oslo, Norway; 2002. xii + 137 p

Section 5

Plankton in Aquatic Ecosystems

Chapter 5

Importance of Plankton to Fish Community

Hamdy Abo-Taleb

Abstract

Zooplankton is the source of life for most of aquatic organisms especially in their larval stages. Its importance come from that most fishes depend on it as a source of life after absorbance of the yolk sac. Moreover, the greatest aquatic creatures like many species of whales are filter feeders where the planktonic organisms form the main bulk of their food. The importance of zooplankton as a main source of nutrition of marine fish larvae has been long professed. Many scientists attributed the ability of a fish population to outdistance through the larval period without vast mortality as one of the primary factors determining the size of the resulting year class and hypothesized that competition for food during the larval time might be a major factor affecting survival and subsequent year class strength.

Keywords: ecology, fish larvae, fisheries, phytoplankton, zooplankton, competition

1. Introduction

The food spectra of the fish family Myctophidae are well-known to be wide, including practically all zooplankton groups, and composed mainly of copepods, euphausiids, and hyperiids, while chaetognaths and decapods are of least importance [1].

Podrazhanskaya [2] studied the feeding habit of the family Myctophidae by investigating a total of 344 stomachs eviscerated from 11 fish species that have been collected from the Northwest Atlantic Ocean; the data revealed that the zooplankton species that occur inside the stomachs of this family with significant biomass were *Calanus finmarchicus*, *Parathemisto norvegica*, followed by *Parathemisto compressa*, then *Metridia lucens*, and some species of the genera *Pleuromamma* and *Conchoecia*. The species *Calanus hyperboreus* and *Metridia longa* (arcto-boreal forms) occurred fairly frequently. Copepods consisted the main food item in the stomachs of *Benthosema glaciale*, *Electrona risso*, *Hygophum benoiti*, *Lobianchia dolfeni*, and *Protomyctophum arcticum*, while *Diaphus rafinesque*, *Myctophum punctatum*, *Notoscopelus bolini*, and *N. elongatus* fed primarily on euphausiids; on the other hand, the dominant food in the stomach content of *Ceratoscopelus maderensis* was the planktonic hyperiids.

It was noticed that the stomachs of *Benthosema glaciale* contained some foraminiferans and algae in addition to the main bulk (copepods), while the predator zooplankton *Sagitta* spp. were recorded in the stomachs of *Electrona risso*. *Tomopteris* (pelagic polychaetes) was found in the stomachs of *N. elongatus* [2].

Boudreau and Dickies [3] established an energy transmit model between contiguous groups in the trophic chain such as zooplankton and fishes based on the following facts:

a. Specific predator biomass is a dependent variable to size.

b. The product of growth efficiency and mortality of the predator imposed on the prey is size vassal.

c. For all neighboring trophic groups, the ratio of the predator to the prey size is constant.

Sprules and Goyke [4] mentioned that *Salvelinus namaycush* (lake trout or large salmon) nourishes on *Alosa pseudoharengus* (smaller alewives) and cisco that utilize a tiny-sized zooplankton as a food. So it is sensible to use the size spectrum of Lake Ontario to prophesy the annual lake production of trout from zooplankton biomass data.

Many researchers suggest that increase in quantity of zooplankton would result in an increase in the quantity of fishes (homeostasis) [5].

The difference in zooplankton production in estuarine, coastal, and oceanic realms has been correlated to the fishery potential of the concerned area. Zooplankton is the main link in the energy transmission at secondary level, they plays a considerable role in the production potency of any aquatic ecosystem. In sum, estimation of zooplankton standing crop gives an index to determine the extent of the sea fertility. To a confirmed extent, the fishery failure and success—specially the pelagic ones—are dependent on the plankton availability. High fish stocks are found in regions of high plankton biomass that in turn are the enrichment regions. In the Indian Ocean and other oceans, there are some reports confirming the direct relation between zooplankton and pelagic fishery production [6–8]; major part of the pelagic fishery is shared by shoaling fishes like sardines, mackerel, etc., which are essentially plankton feeders.

Evaluation of fishery potential is based on the hypothesis that about 10% is the ecological efficiency to transfer from trophic level to the next one. In the ocean food web, around 10% of zooplankton production (secondary production) will be obtainable to fish (tertiary level). Predominantly the validity of such assumption was questioned. Chapman [9] proposed that the value of ecological efficiency is 25%. Around 7.47 is the factor that was used to elevate the value of the carbon to get the fish wet weight [10].

Plankton community can determine or control the fish population; it can assist us in determining (1) seasons and regions of spawning, (2) adult spawners' biomass, (3) adult annual variation (biomass), (4) adult migrations, (5) growth performance and survival rates of larval stages, (6) relation of environmental conditions to abundance and distribution of the mature (adult) and immature (larvae) forms, (7) zooplankton and fish larva trophic relations, and (8) interactions among species throughout larval stage that might thereafter influence stock size.

2. Fish and plankton relationship

What are the reasons that make several fish species superabundant in the sea or why are many types of fish so abundant and successful in the ocean to the degree of attracting human by their abundance? The aquatic biologists can answer this by their broad studies on the ocean environment with assertiveness on the vital plankton role. Plankton has an essential role in the fluctuations happening in the naturalistic survival rates of fish juvenile and larvae and the consequent effects on the adult fish stock. In another mean, tiny organisms control the development of the future fish stock. As larvae grow, many become less dependent on plankton, thereby reducing mortality rates with age [11].

Fish growth and mortality can vary significantly under normal conditions. The laying of expansive number of eggs from which just a few can survive and grow up to become adults is a typical r-strategy. The larva success will be decided by its surrounding environmental conditions and plankton abundance and not by the quantity of parent stock. At many lower survival levels, there is a density reliance where the amount of the available food to each individual plays a fundamental role in determining how many will survive additionally, the production of live food is related to fish nourishment type and it is also a densityindependent operation. Ordinarily spawning sites and seasons were determined by checking the gonad status of fishes held at different times and areas all year round. In any case, later information on component of ichthyoplankton dependent on the real spawned eggs and larvae gathered have supplied us with a full picture of the spawning seasons and sites. Additionally this information has given us the knowledge of fecundity per unit weight and the female's proportion in the fish stock [11].

The following points are worthy of consideration and are consistent with Balachandran and Peter [11]:

- Oceans' fertility is measured by plankton biomass which is considered an index on it. It supplies us with an estimation of the total organic production and helps us to chart out the sites of fishery potential.

- The production of fish can change as a result of the productivity changes with which the plankton is converted into the tertiary production (fishes) rather than the total primary production changes.

- In oceans, plankton are the essential source of food; the inconstancy in their composition (their diversity) influences the fishes' food habits.

- Plankton community structure indicated the central role of such organisms as a vital factor in the fishes spawning.

- A fishery survives and becomes wealthy when a mix of favorable conditions prevails that causes the food to be supplied in adequate amounts and decreased prey density.

- One of the main factors determining the size of the resulting fish stock year class is its capability to exceed the larval stage without massive mortality.

- Predation of zooplankton on fish larval stages influences the following year class quality/fisheries.

- It is important to determine the starved larva percentage which can be used as a signal of eventual year class strength.

- Plankton bioindicator concept means using of certain species of plankton as indicators on the fishery status. Nowadays it is extremely needed as well as very applicable.

3. Fishery management

Plankton play an essential role in management of fisheries that can be as per the following:

A. Size of the spawning stock can be directly gauged from the survey data of the egg and larvae, and when matched to statistics of particular fisheries, catch can be used to determine when the overexploitation level is being near.

B. The traditional method of assessing stock sizes for commercial fisheries was found unsuitable where fisheries nowadays have been unauthorized because of stock depletion.

C. Plankton studies help us to comprehend the natural aquatic ecosystem with a view to answer the prediction question: what amount of fish can be gotten from the ocean?

D. Studies on plankton allow us to grasp the impact of catching large amounts of fishes from the same natural environment.

4. Larval abundance and plankton

There is a correlation between abundance of fish and the plankton abundance. Fishery wealth is commonly depending on the wealth of plankton. The mechanism of depensatory can be obviously noted when a great number of larvae contest so vigorously for a limited food amount that the survivors may turn out to be a more fewer than if less larvae compete initially. The greater the competitors, the fewer the survivors and vice versa [11]. Often predation by planktivorous fish on plankton community may affect the size structure of an aquatic food web [12]. Abundance of planktivorous fish promotes growth of smaller zooplankton and phytoplankters, while larger zooplankton flourished with fewer fishes.

Peter [13] studied the volume of the plankton samples and their relationship to the respective numbers of fish larvae and eggs in the Bay of Bengal and the Arabian Sea. The author found inconsistent relationship. Furthermore an inverse relationship was recorded in some cases. These can be attributed to the samplers used, avoidance of net by larvae, and the nature of sampling (vertical hauls were made instead of oblique hauls). On the other hand, Devi [14] noticed no relationship between the plankton volume and the larva numbers. Identical records were noted by many authors such as [15–18], while George [19] has mentioned that there was a positive correlation in southwest coast of India at the coastal waters. Studies on zooplankton showed a very general relationship between zooplankton and fish larval abundance.

The importance of zooplankton as a principal source of food of marine fish larvae has been long recognized. At the beginning of the nineteenth century, Hjort [20] imputes the fish population ability to exceed the larval time without massive mortality as one of the essential factors determining the future size of the output year class. Saville [21] assumed that the food competition during the period of the larva might be the main factor influencing survival rate and the subsequent strength of year class.

According to Holt and Beverton [22] and Ricker [23], the stock/recruitment relationships proved that the larvae survival can be density-dependent especially at high stock densities because recruitment does not accelerate at levels of high stock. Factors of density dependence may run at either the inter- or intraspecies level through the larval stages. Relatively 2 days after hatching or after yolk reserves exhaustion, the fish larvae become ready to feed as soon as possible.

Hunter [24] mentioned that there are many reasons that can cause larval mortality; the main reason is the larval starvation after absorption of their yolk sac. There are several measurements to detect the larva starvation which include morphology of the larva, morphometric measurements, histological investigations, and chemical analysis.

Larval feeding is effected by several factors including the extent of availability of suitable and sufficient food at a suitable concentration.

5. Fish larvae and plankton

5.1 Predation relationship

Looking at plankton, not as a food for the fish larvae, but as a predators show another aspect of those organisms. The analyzed collections of plankton so far confirmed the existence of carnivorous predator species such as chaetognaths, chondrophores, ctenophores, medusae, and siphonophores sometimes in a great number. Chordates, copepods, decapod larvae, euphausiids, heteropods, pteropods, and polychaetes existed in high abundance in some of the investigated collections.

The environmental factors more or less controlled the abundance of the previously mentioned species. Considerably a great number of different fish larvae in various digestion stages were noticed in the guts of these predator plankton groups.

Comparing with relatively sluggish yolk-sac-bearing larvae, the previous predators preferred the active swimmer fish larvae. As ctenophores drifted at the sea surface or subsurface, their predatory behavior is restricted to this zone, while others like siphonophores are the most active predators as they could swim quickly through the water column. The predatory efficiency of an organism is strongly correlated with its size, while the vulnerability of the prey (any species) is associated with its abundance and size. Around 108 predator plankton species were recorded by Alvarino [25]; she recorded their maximum abundance in hauls where there are no anchovy larvae; in those with aggregations of larvae, she found domination of copepods and/or euphausiids, and larvae were missed in hauls which are predominated by pelagic protochordates.

Predominantly the zooplankters' predatory pressure is being weaker, when there is a high density of copepods which could appear by the analysis of gut content of these zooplankter species. There are several recorded cases of heavy predation on fish larvae by different plankton groups such as chaetognaths and ostracods which were recorded by Lebour [26] and Nellen [27] or by medusa like *Cyanea* and *Aurelia* [28], and the common predation by copepods [29] was also reported. Predation activity can be annually varying and consequently affecting the subsequent year's class strength.

Hunter [24] mentioned that in a recent academic conference related with the mortality of fish larvae, it was summarized that "starvation and predation are the main reasons of larval mortality, and to some extent these two may interact." It is a very well-known fact that larval mortality as a result of predation is decreased when potential predators are lesser in number, and mortality caused by starvation is decreased when fish larvae are found in waters with a suitable food supply. A fisheries enhanced and survives when an appropriate group of factors as above are dominates "e.g. food abundance, predators decreasing, ...etc" as in the previous noted case "water of anchovy".

Predators eat a huge number of fish larvae which may explain the great larval mortality. Juveniles formed larger prey items than the eggs but less in numbers. Eating fish larvae for large numbers of plankton leads to faster growth and vice versa. Feeding is necessary for survival of larvae, while it is necessary to increase the growth rates of juveniles, all related with the availability of plankton.

For the planktonologist, the most essential aspect of fish and larvae growth is not only the abundance or availability of any prey or food items but also the presence of the right food type [30], as the growth performance is correlated to the nature and type of food [31]. Leong and O'Counell [32] observed rate of feeding alteration by anchovies by changing the method of prey catching from filter to raptorial nutrition.

5.2 Plankton as indicators of fishery

Plankto-trophic larva is a stage characterized by majority of fishes, their existence indicating the presence of the adult stages which shape the fishery. They also act as bioindicators, whose larval life is prolonged as flat fishes. Nair [33] and Nair and Subrahmanyam [34] linked the fluctuations in the oil sardines with the existence and flourishing of a diatom *Fragilaria oceanica*. Selvakumar [35] found a relation between mackerel fisheries' wealth and blooming of cladocerans *Evadne* and *Penilia*. Sakthivel [36] has commended on the pivotal role played by pteropods as bioindicators and as food for tuna and herring. Alvarino [25] related occurrence of *Sagitta decipiens* along with anchovy larvae.

5.3 Food chain relations

Determining the primary production as well as the quantitative transfer among trophic levels, the assumed production of fish in an area, both in the early stage "zooplankton-eater larvae" and later predators, also can be evaluated with a great importance. As indicated by Gulland [37], contrasted with the all-out yearly primary production in the seas of about 20×10^9 tons of synthesized carbon, the fish catch amounting to 5×10^6 tons of carbon (100×10^6 tons/year of fish) demonstrates a distinction of 4000 multiply. This is on the grounds that the fish being fished are different stages expelled from the primary production experiencing about 90% decrease at various trophic levels.

A report on the data of the world oceans' primary production was recorded by Koblentz-Mishke et al. [38]; the authors observed that the primary production over vast regions of seas was notably low, while productivities were higher, i.e., 2–3 times more, in the closeness of land masses. Platt and Subba Rao's [39] introduced synopsis shows that the total primary production in the world seas is around 31 × 109 tons of carbon every year. This information clearly warrants a crisp taking a gander at past estimates.

5.4 Trophic levels

It is enjoyable to memo that certain species get transferred from one level to another as they grow. We can classify the trophic levels into the following classes as agreed with Petipa et al. [40]:

1. The trophic levels started with autotrophs and saprophages that lie at the first trophic level.

2. The second trophic level includes herbivorous organisms such as copepod nauplii, many copepodite stages, *Oikopleura* spp., and the larvae of many polychaetes and benthic molluscs.

3. At the third level there are the omnivorous fauna which include the premature later-stages of some copepods such as *Acartia* spp., *Oithona* spp., and *Centropages* spp.

4. The fourth level contains the primary carnivores as many adult copepods like *Oithona* spp.

5. The fifth level involves the secondary carnivores like *chaetognaths*.

6. The sixth level comprises the tertiary carnivore as *Pleurobrachia* which nourish on all other zooplankton species.

On the other hand, Ryther [41] divided the trophic levels into three food chains that depend on the different communities which can be described as follows: (1) oceanic level, (2) continental shelf level, and (3) upwelled community level.

1. Oceanic level: Ryther propose that the communities that were known as oceanic have long food chains as an uninterrupted flux of biomass from phytoplankton till fish, with low ecological efficiencies determined by the three or four carnivorous feeding levels. The primary production of the oceanic region mostly slows down annual average value of 50 g carbon/m^2/year which required five trophic levels reaching to production of fish.

2. The second type of food chain "continental shelf or coastal" is found in regions where about 100 g carbon/m^2/year is the average annual primary production; it is composed of three trophic levels whether this is through the pelagic or benthic community.

3. The third chain is upwelling zones with 300 g carbon/m^2/year annual primary production which consists of one and a half trophic level whales feeding directly on euphausiids and adult anchovy feeding immediately on phytoplankton.

A change in the pattern of feeding can reduce the overall efficiency of transfer of energy. The changes in the efficiency may be due to qualitative changes in the zooplankton consumed as in the case with herring feeding on large *Calanus* instead of small *Temora* and *Pseudocalanus* [42].

Plankton have direct and indirect impact on the fisheries' healthiness as they are considered a direct food for some planktivorous fish [43, 44] or because phytoplankton and zooplankton development (plankton-rich water) leads to the flourishment of the fouling community.

Author details

Hamdy Abo-Taleb
Faculty of Science, Al-Azhar University, Cairo, Egypt

*Address all correspondence to: voalia@yahoo.com

IntechOpen

© 2019 The Author(s). Licensee IntechOpen. This chapter is distributed under the terms of the Creative Commons Attribution License (http://creativecommons.org/licenses/by/3.0), which permits unrestricted use, distribution, and reproduction in any medium, provided the original work is properly cited.

References

[1] Vinogradov ME. The Vertical Distribution of the Oceanic Zooplankton. Moscow: Nauka; 1968. 357 p

[2] Podrazhanskaya SG. Feeding habits of mesopelagic species of fish and estimation of plankton graze in the Northwest Atlantic. NAFO Scientific Council Studies. 1993;**19**:79-85

[3] Boudreau PR, Dickie LM. Biomass spectra of aquatic ecosystems in relation to fisheries yield. Canadian Journal of Fisheries and Aquatic Sciences. 1992;**49**:1528-1538

[4] Sprules WG, Goyke AP. Size-based structure and production in the pelagia of lakes Ontario and Michigan. Canadian Journal of Fisheries and Aquatic Sciences. 1994;**51**:2603-2611

[5] Leukowicz M. The communities of zooplankton in fish ponds. Acta Hydrobiologica. 1978;**16**:139-172

[6] Sudarson DJ. Observations on the plankton and trawler catches off Bombay. Marine Biological Association of India. 1964;**6**(2):222-225

[7] Prasad RR. Zooplankton biomass in the Arabian Sea and the bay of Bengal with a discussion on the fisheries of the region. Proceedings of the National Academy of Sciences, India. 1969;**50**:115-252

[8] Nair VR, Peter G, Paulinose VT. Zooplankton studies in the Indian Ocean II, from the Arabian Sea during the postmonsoon period. Mahasagar-bulletin of the National Institute of Oceanography. 1978;**11**(1&2):35-43

[9] Chapman WM. Potential resources of the oceans. In: Camp V, editor. World Food Problem. Port of Long Beach. California, 43 PP. Oliver Ef Boyd: Van Camp Sea Food Co., Edinburg; 1965. pp. 142-167

[10] Cushing DN. Production in the Indian Ocean and the transfer from the primary to the secondary level. In: Zeitschel, editor. The Biology of the Indian Ocean. Ecological Studies. Berlin: Springer-Veriag; 1973. pp. 475-486

[11] Balachandran T, Peter KJ. The role of plankton research in fisheries development. Bulletin of Central Marine Fisheries Research Institute. 1989;**44**(1):163-173

[12] Brooks JL, Dobson SI. Predation body size and composition of plankton. Science. 1965;**150**:28-35

[13] Peter KJ. Studies on some fish larvae of the Arabian Sea and Bay of Bengal [thesis]. University of Cochin; 1982

[14] Devi CBL. Studies on the flat fish (Heterosomata) larvae, of the Indian Ocean [thesis]. University of Kerala; 1936

[15] Strasberg DW. Estimates of larval tuna abundance in the Central Pacific, 1960. United State Fishery Bulletin. 1967;**50**:231-235

[16] Nakamura EL, Matsumoto WM. Distribution of larval tunas in Marquesan waters. United State Fishery Bulletin. 1966;**66**(1):1-22

[17] Alikhan J. Distribution and abundance of fish larvae in the Gulf of Aden and in the waters off the coast of W. Pakistan in relation to the environment. Diss-Kiel. 1972;**1**:191

[18] Ashour M, Abo-Taleb HA, Abou-Mahmoud MM, El-Feky MMM. Effect of the integration between plankton natural productivity and environmental assessment of irrigation water, El-Mahmoudia Canal, on aquaculture potential of *Oreochromis niloticus*. Turkish Journal of Fisheries and Aquatic Sciences. 2018;**18**:1163-1175

[19] George KC. Studies on the distribution and abundance of fish eggs and larvae off the southwest coast of India with special reference to Scombroids [thesis]. University of Cochin; 1979

[20] Hjort J. Fluctuations in the great fisheries of northern Europe viewed in the light of biological research. Rapports et Proces-verbaux des Réunions. Conseil International pour l'Éxploration de la Mer. 1914;**20**:1-228

[21] Saville A. Application of Ichthyo-plankton studies of fishery management. UNESCO Technical Papers in Marine Science. 1975;**20**:25-27

[22] Holt SJ, RJH B. On the dynamics of exploited fish populations. In: Fishery investigations Series II. London: Ministry of Agriculture, Fisheries and Food, Great Britain. Vol. 19; 1957. pp. 1-533

[23] Ricker WE. Handbook of computation for biological statistics of fish populations. Journal of the Fisheries Research Board of Canada; 1958. pp. 119-300

[24] Hunter JR. Report of a colloquium on larval fish mortality studies and their relation to fishery research January, 1975. NOAA Tech. Rep. NMFS dr.; 1976. p. 395

[25] Alvarino A. The relation between the distribution of zooplankton predators and anchovy larvae. Rapports et Proces-verbaux des Réunions. Conseil International pour l'Éxploration de la Mer. 1981;**178**:197-199

[26] Lebour MV. The food of plankton organisms. Journal of the Marine Biological Association of the United Kingdom. 1923;**12**(4):644-677

[27] Nellen W. Kinds and abundance of fish larvae in the Arabian Sea and Persian gulf. In: Zeitzschel B, editor. The Biology of the Indian Ocean. Ecological Studies. Berlin: Springer-Veriag; 1973. pp. 415-430

[28] Fraser JH. Experimental feeding of some medusae and chaetognatha. Journal of the Fisheries Research Board of Canada. 1969;**26**:1743-1762

[29] Lillelund K, Lasker R. Laboratory studies of predation by marine copepods on fish larvae. Fishery Bulletin. 1971;**69**(3):655-667

[30] Parsons TR, Lebrasseur RJ. The availability of food to different trophic levels in the marine food chain. In: Steele H, editor. Marina Food Chains. Edinburg: Oliver & Boyd; 1970. pp. 325-343

[31] Paloheimo JE, Dickie LM. Food and growth of fishes. III. Relation among food, body size and growth efficiency. Journal of the Fisheries Research Board of Canada. 1966;**23**:1209-1248

[32] Leong RJH, O'Connell CP. A laboratory study, of particulate and filter feeding of northern Anchovy (*Engraulis mordax*). Journal of the Fisheries Research Board of Canada. 1969;**26**:657-582

[33] Nair RV. Studies on the fife-history' bionomics and fishery of the white sardine, Kowala coval (Cuv.). Proceedings of Indo-Pacific Fisheries Council. 1953;**20**:103-118

[34] Nair RV, Subrahmanyam R. The diatom, Fragilaria Oceanica cleve, an indicator of abundance of the Indian oil sardine, Sardinella longiceps Cuv. & Val. Current Science. 1955;**24**:41-42

[35] Selvakumar RA. Cladocera swarm in relation to mackerel fishery along the west coast of India. Current Science. 1970;**39**:481-482

[36] Sakthivel M. Studies on the Euthecosomata of the Indian Ocean [thesis]. University of Cochin; 1972

[37] Gulland JA. Food chain studies and some problems in world fisheries. In:

Steele H, editor. Marina Food Chains. Edinburg: Oliver & Boyd; 1970. pp. 296-315

[38] Koblentz-Mishke OJ, Volkovinshy VV, Kabanova JG. Plankton primary production of the world oceans. In: Wooster WS, editor. Scientific Exploration of the South Pacific. Standard Book No. 309 01755-6. Washington: National Academic of Science; 1970. pp. 183-193

[39] Platt T, Subba Rao DV. Primary production of marine microphytes. In: Cooper JP, editor. Photosynthesis and Productivity, in Different Environments. Washington: National Academy of Science; 1975. pp. 249-280

[40] Petipa TS, Pavlova EV, Midonov GN. The food web structure, utilization and transport of energy by trophic levels in the planktonic communities. In: Steele H, editor. Marine Food Chains. Edinburg: Oliver & Boyd; 1970. pp. 142-167

[41] Ryther JH. Photosynthesis and fish production in the sea. The production of organic matter and its conversion to higher forms of life vary throughout the world ocean. Science. 1969;**166**:72-76

[42] Steele JH. Some problems in the study of marine resources. Special Publications International Commission for North-West Atlantic Fisheries. 1965;**6**:463-476

[43] Abdel Aziz NE, Aboul Ezz SM, Abou Zaid MM, Abo-Taleb HA. Temporal and spatial dynamics of rotifers in the Rosetta estuary. Egyptian Journal of Aquatic Research ,Egypt. 2011;**37**(1):59-70

[44] El-Feky MMM, Alprol AE, Heneash AMM, Abo-Taleb HA, Omer MY. Evaluation of water quality and plankton for Mahmoudia Canal in Northern West of Egypt. Egyptian Journal of Aquatic Biology & Fisheries. 2018;**22**(5):461-474